DOWNWARD
SIZING
DOG

A REFORMED BIG DOG SNOB
DEFENDS THE SMALL DOG LIFE

DOWNWARD
SIZING
DOG

A REFORMED BIG DOG SNOB
DEFENDS THE SMALL DOG LIFE

KAREN LENA IZZO

SMALL DOG
PRESS

©Copyright 2023 Karen L. Izzo

All Rights Reserved. No part of this publication may be reproduced, distributed, or transmitted in any form or by any means, including photocopying, recording, or other electronic or mechanical methods, without the prior written permission from the author, except in the case of brief quotations embodied in critical reviews and certain other non-commercial uses permitted by copyright law.

Cover Design by Robin E. Vuchnich
Interior Formatting by Tracy Atkins, The BookMakers

ISBN: 979-8-9869234-0-6 (paperback)
ISBN: 979-8-9869234-2-0 (hardcover)
ISBN: 979-8-9869234-1-3 (e-book)

SmallDogRules.com

Small Dog Press, LLC
Warwick, Rhode Island

Disclaimer

Downward Sizing Dog provides generalized information regarding small dogs, and I've made every effort to be accurate and complete in these pages. This book may contain opinions or information that scientific thought later deems inaccurate. I do not provide veterinary, nutritional, or behavioral advice for you to use on your dog. You must always, always, always consult your own knowledge of your dog and your circumstances and then bring in appropriate professionals to assist you in caring for your canine family member.

Dedication

To my parents, Jack and Ginny Crandall,
who fostered my deep connection to nature and
showed me the pleasure of hard work, done well.

Contents

Dedication iii
Contents v

Prologue 1

Chapter One: Big Dog Ego 7
   Needs and Wants 12

**Chapter Two: Run With The Big Dogs,
Or Stay On The Porch** 16

Chapter Three: Small Dogs Are Dogs 24
   Small Dog Genetics 25
   Small Working Dogs 31
   Small Dog Heroics 36
   Small Dogs as Medical Detectives 39
   Small Dogs Outdoors 40

Chapter Four: Small Dogs Are Smart 45
   Early Human Perceptions of a Dog's Mind 48
   The Scientific Renaissance View of a Dog's Mind 49
   Victorian Evolution: Darwin 51
   The Dawn of Dog Cognition in the Twentieth Century 52
   The Twenty-First Century: Every Dog DOES Have Their Day 54
   And Now the Main Event: The Science of Small Dog Smarts 54
   More Stories of Small Dog Smarts 61

Chapter Five: Small Dogs Are Powerful Health Boosters .. 73
   Small Dogs Ease Situational Stress 75
   Small Dogs Aid Sleep 78
   Small Dogs Improve Our Mental and Physical Health 79
   Small Dogs at Work 91

Chapter Six: Small Dogs Are Easier on Our Allergies, Our Homes, and Our Planet **99**
   Dogs and Human Allergies 99
   Small Dogs Equal Cleaner Homes 102
   When Big Dogs Make Sense 104
   When Small Dogs Make More Sense 105
   Small Dogs Are Easier on the Yard 108
   Small Dogs Are Green(er) 109

Chapter Seven: Small Dogs Travel Well **113**
   Have Small Dog, Will Travel (by Car) 114
   Have Small Dog, Will Travel (by Plane) 119
   Getting from the Airport to the Hotel and Everywhere Else 123
   The Traveling Welcome Mat: Small Dog-Friendly Stays . 124
   When Do We Leave? 126
   Small Dogs Have Built-In Babysitters if They Must Stay Home 126

Chapter Eight: Small Dogs Are Less Expense **129**

Chapter Nine: Small Dogs Live Longer **138**

Chapter Ten: The Small Dog Life Fits Every Life Stage **142**
   The Case for Small Dogs and Kids...................................... 144
   The Case for Small Dogs and Millennials 154
   The Case for Small Dogs and Boomers 155
   The Case for Small Dogs and Pet Parents in Late Seventies and Beyond..160
   Illness, Aging, and Small Dogs 166
   At Every Age: Emergency and Estate Planning for Small Dogs 174

Chapter Eleven: But Wait, What About Small Dog Syndrome? **181**
   The Vaccine for Small Dog Syndrome 185
   Small Dog Culture.. 187
   House-training..190

Chapter Twelve: A Brief Bit of Advice About Finding Your Small Dog 205
Avoiding Small Breed Puppy Mills 206
Understanding Breed Groups, and the Smallest Dogs in Each 209
The One Thing 217

Epilogue: The Small Dog Rules 220

Acknowledgments 223

About the Author 225

Downward Sizing Dog: The Companion E-Book 226

Endnotes 227

Prologue

*C*laire and Tess, classically gorgeous English setters, stand nearly hip high and weigh roughly seventy pounds apiece. By any standard, they're big and they're beautiful. They also consider themselves lap dogs, and any human's lap will do. They give bonus points if your lap is anywhere in the vicinity of snacks. They're particularly fond of butter.

As a hunting breed, they require significant daily exercise, or they're not happy. On the rare occasion they miss their daily workouts, they freely share their dissatisfaction, ramping up their restless movements through the house. In one instance of particular indignance when houseguests disrupted their routine, Claire leapt from the chair to the sofa to the chair again, knocking over a lamp in the process. In her defense, some rough-housing grandchildren might have helped spur the chaos, but still.

As avid outdoor enthusiasts, Claire and Tess's owners take them on daily four-miles treks to nearby Wakeley Lake, where both dogs run long and free. Or, when it's too hot to hike, they all clamber into the beautiful Au Sable River that rambles past the house, the pups swimming upriver against the relentless current. Not surprisingly, Claire and Tess are incredibly calm, fit, and happy. So are their owners, Jack and Sandi Crandall. They're my dad and his wife, ages eighty and sixty-two.

DOWNWARD SIZING DOG

My dad and Sandi have since had to say goodbye to Claire, who, like most of my dad's hunting dogs, lived into her mid-teens. Tess is now twelve, and her four-mile walks occur only every other day, as her joints don't cooperate with the daily marathons that once sustained her. Swimming still happens nearly every summer afternoon. In winter, her off days still find her outside for hours, working side-by-side with her dad (and mine) as he beats the snow into submission, fills their massive bird feeding platforms, trims errant tree limbs, and generally savors their favorite season.

Dad and Sandi live a life ideally suited to joyful big dog lives: They enjoy vast outdoor space, ample time, and an innate love of the woods, rivers, and lakes of Michigan. Their animals are fundamental to their daily rhythms, and these large sporting breeds have always been essential companions in the hunting, hiking, and outdoor adventuring that have long fueled them all. Their dogs live life to the lees. So, the entire premise of this book notwithstanding, I freely admit big dogs are indeed the right fit for some of us.

In fact, I wouldn't even be talking about little dogs if it weren't for all the big dogs who came first in my life. Yet, although big dogs might be right for some of us, small dogs are probably best for many of our modern lives. As we explore throughout this book, downward sizing our dogs can be a smart thing for us and for our dogs. It's not settling for less, and small dogs are not some inferior creatures, destined to come in a distant second to powerful sporting or working breeds. Fortunately, many of us are realizing that, as did I, one fall morning nearly eight years ago.

I smile at the man, who, under other circumstances, would be someone I'd rather not encounter in a dark alley.

Prologue

It's a blustery but warm late autumn morning, perfect for a seaside walk, and I slow my pace as I near him. He's leaning against the Narragansett sea wall in worn jeans, his black T-shirt gracing the ink on his weathered arms. His muscled frame ends in biker boots—a subtle communication to strangers that he's not to be messed with.

But that doesn't stop me.

A director casting him in a B movie would predictably round out his character with a typecast pittie, or perhaps some type of police dog—a German shepherd or Doberman. Instead, the man is cradling a soft, beige ball, shielding the bundle from the whipping sand that swirls around us in baby tornadoes.

A sudden lull in the wind creates a calm.

A curious snout appears, and then eyes peer out from the protective crook of the man's tatted arm.

"Do you catch a lot of grief over that little thing?" I give a head-nod to his pint-sized Yorkie as I stop and wait for his reply.

He grins, nodding "yes" as he skritches the top of his miniature companion's head. I ask permission and then reach out to ruffle the hair under her chin, receiving a friendly lick from her tiny pink tongue in return for my regard.

We stand together for a moment, enjoying a companionable silence that only small dog owners can understand.

It's a knowingness—an unspoken camaraderie that binds us in our shared understanding of the powerful animals packaged in such a convenient size.

My story took place over seven years ago, and the entire encounter lasted only moments. But I love this memory as it marks the day I took a hard look at the vestiges of my big dog ego and vowed not only to walk away from it, but to embrace the beauty of small dog life.

Proudly.

It's also the day *Downward Sizing Dog* found its way into my consciousness.

These moments elevated both my life and my life's path. If a guy in biker boots and skull tattoos could embrace his inner small dog, then so could I. I'd already been experiencing the everyday joy that comes with owning a small breed dog. And when I sat down to research this book, I became a true convert, realizing that *almost* everyone thinking of getting a dog, or another dog, would have a better life if they just squashed their own preconceived notions and went small instead.

If you're suffering from a big dog ego, with its vision of "family" that often makes little sense for either the family or the dog, you'll find better options here. When I stopped feeling defensive about our small dogs, and I instead started sharing stories of their endless capacity to make us laugh, understand our emotions, and improve our physical and mental well-being, I stumbled on what shouldn't be a well-kept secret: Small dogs are a cure for much of what ails us.

They provide the canine connection most humans crave with a fraction of the emotional, physical, environmental, and financial stressors involved in ensuring a big dog's happiness. If you want a dog, a small dog will enrich your life beyond measure. And that little canine will do so in a way that likely fits almost seamlessly in your life, provided you understand the time and care all dogs require.

Indeed, I experience near-daily proof of that beautiful life meshing of human and small dog. Consider the time, for instance, when extended family invited us to a last-minute gathering on the Cape—we were all to meet at a beautiful inn for a summer evening dinner. But it was a commitment, as we faced a two-hour drive there during height-of-summer traffic. And the drive back is always worse.

The problem? There was none.

When we had two golden retrievers, we would have declined the invitation, as our dog sitter and every human we know would be otherwise booked, basking in the glorious summer weather. But with our shin-high, non-shedding, well-behaved Havanese Phoebe and Scout? One quick text to our hostess, and she's happy to let the girls stay at her home when we all go to dinner. You just can't ask people such favors when what you'd be inflicting on them is one hundred and fifty pounds of golden retriever. And that's just the hair they shed. That's an ask of a major magnitude, akin to expecting someone to take you to the airport at four in the morning.

You just don't do it.

With small dogs, though, you're usually good to go. In the chapters that follow, we'll deconstruct and destroy negative small dog myths, and we'll delve into the many ways small dogs can provide ideal companionship. Small dogs are smart, clean, and green. Small dogs travel big and live long. Small dogs are better for our bodies and (somewhat) easier on our wallets. Small dogs are as intelligent, as empathetic, and as entertaining as big dogs. Here, you'll discover how small dogs can transform and better the lives of the humans who love them.

And that's what this book is about. It's about taking a hard look at the stereotypes that drive us to choose our dogs for the wrong reasons. And it's about walking into a more cohesive life with small dogs who meld beautifully into almost everything we do.

CHAPTER ONE

BIG DOG EGO

I feel his eyes on me, but I ignore the prickly intrusion and continue studying the claw hammers on the wall. Still, his gaze persists, and I intensify my mental command: "Just walk on by, dude."

Dude doesn't obey. Instead, he focuses on the slender leather strap dangling at my wrist, and his glance drops to the floor.

Well, not really the floor.

It would be more accurate to say eight inches off the floor.

We're all standing in the hardware aisle at Home Depot, and at the other end of the leash my Havanese puppy Phoebe embraces his approach, her tail wagging hard, as though she's trying to prove the tail can, indeed, wag the dog. She's freshly groomed, with ridiculous pink neon ribbons fastened to each ear. The bright fabric wobbles like garish devil's horns as her entire body wriggles in welcome, her enthusiastic greeting encouraging our stalker to bend down with a meaty hand to pat Phoebe's head.

I feel that familiar, warm rush of embarrassment that accompanies someone sizing me up based on my choice of miniature canine companion. Anticipating the predictable comments about "yappy little dogs," I attempt preempting him, declaring: "I have an eighty-pound Doberman at home."

He parries: "Well, whatcha' got here belongs in the cleaning aisle. It's a five-pound Swiffer."

He plays fast and loose with his syllables and the word "here" comes out "he-ERR."

I squelch the urge to tell him he looks like he should spend some time in the cleaning aisle himself and instead respond: "She's not five pounds. She's a puppy, and she's eight pounds." My defensiveness hangs in the air, and I add, "She's very tough. She beats up on my Doberman all the time."

He snorts. "Well, then your Dobie must not be much of a Dobie."

That's pretty true, I admit silently.

My Dobie is, indeed, not much of a Dobie. Because honestly, there isn't any Dobie at all. I made that part up. It sounded tougher than the truth.

I do have a ninety-pound dog at home. Well, not ninety. But eighty-five. Almost eighty-five. He's a gorgeous blond, block-headed golden retriever named Romeo.

Romeo's at home because I haven't yet mastered controlling both him and my "Swiffer" Phoebe, who, standing a fierce shin-high, is ready to take on all typical home improvement store threats—the sparrow flitting among the rafters, the water puddling in the garden section, and the German shepherd sounding a low, warning bark in the big truck parked near the garden center.

Plus, with Romeo's bulk, sneaking into stores isn't an option. When he goes, he goes big. And so, although I've abandoned him at home, I feel no guilt claiming he's an imaginary Dobie and parading him out—in conversation, anyway—as a badge of honor.

Because, as everyone knows, there are two kinds of dog people: big dog people who live life with relish and strength, and small dog people who, well, might as well have cats.[a]

That, anyway, was my core belief since childhood.

Big dogs have been part of my life since my earliest memories. Jill—my dad's first hunting dog—was a beautiful Brittany. My mom and dad were in their mid-twenties with a baby, four-year-old me, and a miniscule house.

It was the 1960s, when theory (manly hunting theory, anyway) dictated that hunting dogs belonged in outdoor kennels—it allegedly conditioned them for cold, grueling days flushing game. So, my dad built Jill a large kennel with a warm doghouse in our backyard and left strict instructions that good bird dogs belong outside much of the time. For reasons unrelated to hunting psychology, my mom happily embraced the dog-outdoor theory too.

Jill, however, did not.

We arrived home one night to find Jill had slipped her collar and escaped her kennel. We never found her. After

---

[a] To my cat loving friends, please forgive my shameless cheap shots at these amazing creatures. I actually like cats. A lot.

Jill, my dad no longer took advice from his hunting dog magazines.

Jill was the last dog we would ever lose. From then on, our big dogs were family. Not only were they *not* kept in outdoor kennels, they took up more than half our beds. Two years after Jill ran away, my dad came home with an unusual choice for a hunting companion. She was a beautiful Irish setter named Meg, and she was a lanky, dark-red creature of determination. She listened to only my dad and then, eventually, me. Meg knew my dad was the boss in the house, and my childish pride swelled when she deemed me second-in-command.

That's when my big dog ego started.

Meg was my frequent companion on summer treks out to the old orchard or the cornfield. She was with me when my dad let me fire a shotgun for the first time, and she always obeyed me even when she didn't want to. She taught me that leadership means both earning trust and standing one's ground.

Meg was the big dog of my childhood and adolescence, and she molded my perception of "dog." My near constant interaction with other impressive canines cemented that perception. Most of my parents' friends had hunting dogs, so we were rarely without the company of big black Labs or setters. I learned by osmosis that a dog is a companion, a protector, and a fellow naturalist. She must track birds and flush them out, and she should look elegant on point. She must run all day like a wild wolf, and then at sundown, lie protectively with the kids on the floor by the fire. A real dog stands skull-high at your hip, the tips of your fingers resting on her head, no bending down required.

And my childhood dogs fulfilled that vision for me. We had access to expansive acres of farmland and forests, and my parents were committed naturalists. Our lives revolved around the outdoors and around our dogs. We and our pups lived the fantasy most of America still buys into—that big dogs mark us as rugged individualists who are one with nature. In this worldview, small dogs are some lesser level of creature, hovering just above "cat" in the animal kingdom.

So, to me, small dogs just didn't count as "dogs."

Given the canine population in this country, much of America agrees with that sentiment. Maybe, like me, you're also drawn to the solidity, grace, and protectiveness of a large, rough-hewn canine companion. Certainly, the American Kennel Club's recent lists of most popular breeds proves that most of us love big dogs and we cannot lie: For years now, most of the top ten favorite dogs boast an average weight of well over sixty pounds and include Labradors, Rotties, German short-haired pointers, and German shepherds. Golden retrievers, my former big dog drug of choice, are usually in the top five on that list. Thus, from my mid-twenties through my mid-fifties, golden retrievers were the dogs who owned my heart.

Although I loved my goldens with fierce adoration, looking back, I can now see I did so in the same way one might love someone who's not especially good for you—or for them. We adored one another and yet we fit rather awkwardly into one another's lives. What made them happy—playing for hours in the snow, or sitting in our laps while blowing their coats, or starting an impromptu wrestling match in my kitchen—often interfered with our happiness. The result? None of us really experienced our best lives.

So, at times, these big, golden beasts wore down my sanity and tore up my home.

And yet still, decade after decade, I clung to the regal canine fantasy in my head, despite the big dog reality in my life: gouged wood floors, punctured leather car seats, knocked-over vases, slobbery woodwork, stolen chocolate cakes, land-mined lawns, late-night skunk encounters, fur-covered everything, ripped clothing, and various broken screen doors. Oh, and did I mention the empty wallet?

Despite the chaos they often made in and of our lives, our goldens were family. Having loved them and known them as amazing beings, I wouldn't take back those relationships for anything.

But had I known then what I know now, I one-hundred-and-ten percent would have chosen differently.

I would have rounded out our family with loyal, funny, athletic, affectionate, intuitive, empathetic beings in the form of small, non-shedding dogs. And our lives would have been so much calmer, cleaner, and fuller. And, so much more content—both on the human side and the dog side. That contentment—along with a deep, joyful journey with two amazing twelve-pound canines—is what we have now.

Needs and Wants

Like all journeys, this one encountered hiccups. As my opening story reveals, I didn't magically shed my pride once I realized small dogs were actually, well, *dogs*. My initial experience introducing a small dog into a big dog den was humiliating.

I'll never forget walking into a house full of about twenty people—extended family and long-time friends—who had gathered for a summer party. As with all such parties in our world, nearly everyone had brought their dogs. One of our friends, John, arrived with his handsome black Lab, Tony, whose sweet, puppy ways persisted despite his graying snout. Our other friends of thirty years had brought their yellow Lab mix, rambunctious two-year-old Thor, whom they'd rescued as a puppy. And then we appeared with our Havanese puppy Phoebe, who was then six months old and weighed about as much.

John, who had always disdained little dogs as much as any other self-respecting Italian male, harbored secret affection for "Phebes," and she for him. When they thought no one was looking, they engaged in a snuggling ritual that defied description. But in a crowd of bonded males and a group of family and friends who've paraded around only with big dogs since childhood? Let's just say it's unwise to expect respect when arriving with a dust ball on a leash.

Heads turn. Smirks appear. Someone says, "You're kidding! It's a Kick-Me Dog."

And so it goes. Few big-dog lovers ever see themselves owning and bonding with a toy dog. A certain mythological dignity and strength emanate from a guy in a flannel shirt enjoying his daily walk with his well-trained retriever. You just can't duplicate that marketing image with a four-pound Chihuahua at the other end of the leash.

Google Images agrees.

A quick web search for "man's best friend" will find you knee-deep in pages of gorgeous retrievers, shepherds, and mastiffs. Large, regal dogs sitting next to their human companions, or watching warily as a stranger approaches, still satisfy something primal in us.

We want to know our animals have our backs.

And big dogs make that possible. Or perhaps not just possible, but a reality. We've all heard stories of German shepherd dogs leading families out of burning homes, golden retrievers saving drowning toddlers, and mastiffs facing down would-be home invaders. Scan the Hero Dog Award nominations and it becomes clear: In the lionheart category, big dogs presumably win.

In stark contrast, when we celebrate small dogs, it's often because they endured neglect or abuse and yet don't hate all humanity. As a result, their mere presence and personalities make them easy ambassadors for their breed or rescue association. In that way, they *are* heroes, showing courage in the face of adversity and coming out, not just alive, but embracing the world. But for families wanting a tough dog who can keep the family safe? If we listen to the press and our vanity, small canines are useless. That's what our big dog egos would have us believe. And yet, like all egos, they lead us astray.

As I've said, mine did for years. I once scoffed at purse puppies and anyone who carried them. I rejected most small dog-loving humans as being unworthy of much attention. I didn't believe toy breeds counted as canines or that dogs under a certain size had much—indeed, *anything*—to offer. To me, they were yappy, snappy, annoying, and probably unhouse-trained.

Such thinking is not only grossly misguided, but bad for you and, worse, bad for dogs. In fact, unless you need a large, powerful dog for a particular purpose (physical assistance, bodily protection, or hard labor in dangerous conditions), you don't *need* a big dog at all.

You may desperately think you *want* a big dog, and I understand that. I lived the sometimes-harsh reality of that fantasy throughout my adult life with four decades of gorgeous golden retrievers. And I wouldn't trade my relationship with those big, loving animals. Nor would my kids. But I'd be lying if I said my motives for choosing my goldens were based in logic or that our life with Quincy, then Maddie, then Haly, and finally Romeo, was an easy one. Or, just as important, that their life was ideal. What if I *had* scrutinized my obsession with big dogs and how well those dogs could enjoy the life we were living at the time?

Often, large dogs joining human families end up being the *opposite* of what is best for them and for us. Before we expand our family to include a fur kid, we need to ask ourselves: Do we need a big breed dog, or do we just think we do? And are we rejecting small dogs because of our prejudices that have no basis in reality?

Toy and other small breeds are as skilled and often better than their big breed brothers at the things we want our dogs to do: protect us and love us. Yet much of America sneers at these small animals, proudly hating on every shin-high canine. If we continue embracing the American mythology of our best friend being some powerful wolf ancestor, too many of us will miss the opportunity to enrich our lives with the best kind of canine companionship. If we remain silent as the big dog ego perpetuates the misguided belief that big dogs are better dogs, we will also fuel the shelter loops where large, unruly adolescent dogs often end up. It's time to honor all canines, regardless of size, for their ancient and abiding connection to humans and to recognize that often, downward sizing dog is what's best for everyone.[b]

So, let's take a closer, critical look at what drives our perception that a life well-lived must include an enormous dog at our feet. And then, we'll look at the beautiful realities of small dog life.

---

[b] Let's be very clear here. I'm absolutely not suggesting we trade-in dogs already part of our families. When I talk of downward sizing dog, I'm speaking of instances in which we're adding another dog to our lives.

CHAPTER TWO

RUN WITH THE BIG DOGS, OR STAY ON THE PORCH

Heads-up: This upcoming chapter is a tough one, as it explores what happens when people adopt big dogs they later realize they can't handle, and it delves a bit into the problem of shelter dogs. If you already "get" the problem, you may want to skip right to Chapter Three: Small Dogs are Dogs.

There's an old Southern saying intended to keep lesser humans out of the way of their tougher and more successful counterparts: "Run with the big dogs or stay on the porch."

The phrase is whipped out in instances involving things like fast cars, big tractors, or complex business negotiations. Even though the saying has nothing to do with *actual* canines, I hijacked it for my chapter title, because it expresses the perceived problem with small dogs: Despite occasional feel-good stories of small dog heroics, Americans believe miniature breeds to be pretty useless, unless you're one of those types who wants to sip lemonade and rock on the porch.

It's still a reality: In America, size *does* matter. Tiny house movement aside, too many of us are, either consciously or unconsciously, still big into big cars, big houses, big business.

And big dogs.

As I pointed out previously, for decades upon decades, Labradors, German shepherd dogs, and golden retrievers have dominated the American Kennel Club's (AKC) breed popularity list.[1] The males of these breeds can top out at over eighty-pounds—the average weight of an eleven-year-old boy.[2] For at least the past ten years, big breeds have cornered the market on the top three spots. Recently, the ever-adorable French bulldog has nosed out some larger competition, but at a stout twenty-eight pounds on average, these pups hardly qualify as small.

To be sure, at least some people are seeing the light, as in 2020 and 2021, both the beagle and the dachshund nosed their way into the top ten. And for quite some time after the turn of the century, the Yorkshire terrier held court at number two or three. Yet for the most part, it's the big dogs who have clout.

Why? Why are we so insistent on shoving big dogs into places they may not comfortably fit? What possesses people to adopt what is, in most cases, the canine equivalent of a prepubescent child?

A century ago, we probably had deeply practical reasons. In the 1880s and 1890s, the top dogs, literally and figuratively, were purposeful—we *needed* setters and pointers who could spend days in the field flushing grouse or pointing pheasant, or we required herding dogs out West who kept wolves from the flocks, as their breeders intended.

Is that what drives modern Americans to choose their dogs? Is it some vestigial practicality remaining from our days of working the land and subduing the wild?

Perhaps.

I know full well some of you are saying, "Well, we still need big dogs now." Certainly, many twenty-first-century dogs fulfill utilitarian purposes; modern small-game hunters spend upwards of $600 million annually on their dogs,[3] and dogs working in police roles or search and rescue are so important that one group of researchers proposed developing "canine performance science" in much the same way we've done with top athletes and military personnel.[4]

To a certain extent, Americans still rely on big dogs for deeply practical reasons: We enlist thousands upon thousands of service dogs for those with disabilities or health disorders. We rely on large canines to sniff out drugs, bombs, crooks, cancer, and now COVID. Indeed, since 9/11, we can't keep up with the demand for bomb-sniffing dogs, who beat out every bomb-detecting technology man can invent.[5] But, despite the countless ways we employ our big best friends, for the vast majority of the 78 million dogs in this country, their only job is that of companion.

So that brings us back to the original question: Why do so many everyday Americans insist on big dogs?

Well, maybe we really don't.

The proof is in our shelter systems, where those adorable AKC big breed puppy registrations do nothing to help the never-ending parade of full-grown big dogs churning through local rescues and pounds.

Run With The Big Dogs, Or Stay On The Porch

The AKC puts together its list of most popular breeds based on new registrations from the previous year. Nearly all those registrations involve puppies. Thus, it's illogical to conclude that the AKC's list reveals our most popular dogs. It would be much more accurate to say the list shows us America's most popular *puppies*. And our most popular *dogs*? Well, that's another story. And one where we turn to shelter statistics to reveal what's truly going on with our big dog egos.

While the AKC list's most popular canines are big sporting breeds, several studies show large dogs are least likely to be adopted from shelters and the most likely to be euthanized.[6] And in direct contradiction to the AKC favorites list of big dogs, we have the puzzling fact that regarding shelter adoption, toy dogs were adopted much more frequently than large dogs.[7] Yet, when Americans set out to buy purebred puppies, the big dogs usually dominate the top three spots on the AKC list.

So, what's going on?

Regarding these toy breed shelter adoptions, one researcher suggests small dogs find homes more frequently because of their cuteness factor. As we'll discuss in later chapters, we humans are biologically programmed suckers for an adorable face. And certainly, that penchant for adorableness could explain away the apparent disconnect between the AKC's most popular breeds and the dogs most likely to be adopted. Perhaps we *think* we want big dogs and so we start with a very cute big dog puppy. And then when the puppy becomes as big as an eleven-year-old and even less obedient than our teenager, we have trouble coping.

The shelter population bears out this unfortunate reality. Aside from beagles and the Chihuahua, whose adorableness in the *Legally Blonde* movies and Taco Bell™

commercials unfortunately inspired the purse puppy phenomenon, medium-to-large breed dogs dominate shelter populations. It's been the case for years and it's not getting any better. Look at the national numbers on Petfinder.com, and you'll see that dogs under twenty-five pounds make up only about fifteen percent of the dogs available for adoption. While some rescue organizations claim that the high percentage of large breed dogs in shelters exist merely because such big breeds are more popular, if that were the case, we'd see similar ratios borne out in the shelter populations.

And yet, we don't. As of 2015, this country was pretty evenly split in terms of big dog/small dog ownership, with a little over half of American households owning breeds or mixes under twenty-five pounds.[8] And yet most shelter dogs—eighty-five percent of them right now—weigh much more than that.

Whether it's because fewer people abandon small dogs to shelters, or we adopt more little ones from them, census data from shelter populations proves many Americans find appeal in a tinier canine package. Tragically, though, those numbers also reveal the challenges many families face in living well with big dogs. The obstacles seem so insurmountable that they give up, give in, and their big dogs end up in shelters.

Historically, the abundance of large dogs in shelters wasn't surprising because most people owned dogs well over twenty-five pounds. But then came a size population shift. According to a *Washington Post* article "Tiny Dogs Are Taking Over This Country," small dog homes surpassed large dog homes in 2008 "for the first time ever."[9] In actuality, 2008 may not have been the first time "ever," given that in the decade of the Great Depression, three small-to-medium breed dogs—the Boston terrier, the cocker spaniel, and the fox terrier—often topped the AKC's list of most popular dogs. [10]

The ensuing decades, however, witnessed a steady rise in the size and weight of America's favorite breeds. And then, perhaps because of the so-called Great Recession, the period between 2008 and 2017 again saw a sharp increase in the small dog population in America.[11] In fact, in 2013, the United States population of small dogs surpassed large dogs by three million.[12]

Clearly, for many of us, the small dog *is* the way to go.

Yet small dogs get no respect, and while some people choose their large dogs for the right reason, much of the country chooses breeds based on perception rather than reality.

Scientists who study the human psychology behind why we do what we do, note that we perceive large dogs to be more intelligent and trainable, and we generally view small dogs as "neurotic, intolerant of others, and introverted."[13] This negative perception leads some of us to conclude that small dogs aren't as smart as their larger relatives, even though multiple studies prove that we're misunderstanding measures of intelligence and often measuring the wrong thing. In later chapters, we'll discuss small breed intelligence in-depth, but let's do a quick preview here.

The problem with our view of small dog smarts is that we tend to confuse obedience and willingness-to-please with intelligence. When we do that, then large dogs—most working and sporting dogs—probably win the IQ contest. The fact is that companion or independent working dogs, often on the smaller side, aren't built to obey humans as their primary purpose. We equate that lack of minding one's master to a lack of intelligence, and that's not an accurate correlation.

Thus, one reason Americans are drawn to large dogs is because we value what we incorrectly perceive to be their higher intelligence. But there are other more compelling reasons that drive us toward the wrong dogs for ourselves and our families. In Chapter One, we considered the iconic Ralph Lauren-esque image of the guy in flannel walking his loyal big breed dog. As much as we're loath to admit it, that image, and others like it, are a key factor driving our decisions. And like nearly every instance when our choices stem from a vision constructed for us—whether by Hollywood or mad men agencies—we can count on bitter truth to rip the blinders from our eyes.

In a stunning confirmation of America's attraction to attractiveness—as opposed to practical things like, oh, trainability and health—scientist Stefano Ghirlanda and his colleagues found some of us appear to ignore a dog's substantive qualities in favor of a breed's *social* popularity.

In his research article "Fashion vs. Function in Cultural Evolution: The Case of Dog Breed Popularity,"[14] published in *PloS One*, Ghirlanda's research "found no indication that breeds with more desirable behavior, longer life, or fewer inherited genetic disorders have been more popular than other breeds...." Ironically, Ghirlanda and his team found the opposite to be true: "Breeds with more inherited disorders have been more popular, rather than less popular, suggesting that health considerations have been secondary in the decision to acquire dogs." To be blunt, this study revealed that many of us are more concerned with conforming to social fashion than with the functions of our companion animals.

As part of his research, Dr. Ghirlanda and his colleagues, anthropologist Alberto Acerbi and American psychologist Hal Herzog, compared the AKC's breed popularity data with the timing of movies starring canines in the leading role. And sure enough, they found that for almost a decade following a popular movie featuring a particular breed, that breed's popularity soared, as evidenced by puppies registered with the AKC.

That research revealed what animal rights protectors have long suspected: People often pick their dogs for the wrong reasons. They become caught up in the romantics of the breed—the German shepherd's loyal, protective nature, the Chihuahua's purse-sized convenience, or the dalmatian's beauty. And they don't do their research regarding how such pups will fit into their lives.

But it doesn't have to be that way. We can adopt a dog that's the right size and temperament to fit our lifestyles, instead of bringing home a canine extension of our egos. So, in the coming chapters, as we talk about the advantages of small dogs, let's also talk about the realities of shoving large dogs into what could instead be a beautiful small dog life.

CHAPTER THREE

SMALL DOGS ARE DOGS

It must be said.

Small dogs aren't cats. They aren't some lesser form of canine. They're not descended from some other, non-wolf creature, as science once tried to tell us. Just like their bigger counterparts, small dogs are descended from ancient wolves, and the tiniest teacup toy dog[c] boasts the same canine DNA as a wolfhound, Newfoundland, and Great Dane—dogs roughly two-hundred times its size.

Small dogs are dogs.

But say "small dogs are dogs," and people look at you like you're ridiculous, in that, well, everyone knows they are technically "dogs."

---

[c] Breeding so-called "teacup" versions of standard toy breeds has a long list of ethical concerns (not the least of which involves animal welfare) that, unfortunately, aren't within the scope of our discussion here. If you're thinking of going this small, please do serious investigation and reconsider such a path.

Small Dogs Are Dogs

But too many of us just don't fully believe it. Our big dog egos get in the way. It's an ego that leads some of us to view big dogs as necessary companions to a life well-lived and small dogs as yappy inconveniences.

As I've confessed before, and will have to admit again throughout these chapters, "small dogs aren't really dogs" was my ingrained, unmovable opinion for most of my life. Friends and family can recount times I smiled disdainfully and commented under my breath about people walking "stuffed toys." I'd wonder aloud why they bothered with such useless, high-strung creatures.

And so, in my twenties, and thirties, and forties, I rounded out our young family with large, boisterous golden retrievers whom we thought we couldn't live without. But we shoved our big dogs into lives they didn't fit: not in the postage-stamp city yard we had when the kids were little and not in our two-lawyer, three-child, over-scheduled suburban household.

We were constantly on the go.

And most of the time, going somewhere meant leaving the dogs behind. Even when our destinations were dog-friendly, it was just exhausting to keep up with the hair, the mud, and the mess.

There's a better way to have a dog, at least for a large swath of the population. Almost everything that defines the concept of "dog"—loyalty, affection, adventure, joy, companionship, safety—is packed into every small dog in existence. Small dogs are dogs.

Science says so.

SMALL DOG GENETICS

As repeated genetic studies reveal, all dogs descended from wolves,[15] though to be sure, they may not be the wolves we know and love now. Current science has been bickering for decades over how and where dogs came to be. Some camps theorize modern dogs are more likely common descendents of an extinct Late Pleistocene[d] wolf, whereas others propose either a two- or three-prong melding of grey wolves, coyotes, and red wolves.[16] Other studies delve into the mystery of whether it was human hunter-gatherers or agricultural societies who helped develop the domesticated dog.[17]

While researchers argue about how and when wolves and humans merged paths and domesticated dogs developed, they agree that all dogs, both big and small, comprise the same genetic stuff. Some scientists believe that most *small* dogs have rather ancient origins from extinct wolves in the Middle East[18] and others were so skeptical of that conclusion it compelled them to write an entire rebuttal paper refuting the idea.[19] As we discuss in greater detail below, our current knowledge tells us that a dog's size–how small he is—comes from a set of genetic instructions that suppress growth. And, from a genetic standpoint, certain small dogs share more in common with large dogs than they do with one another.

In fact, it surprised me to find data suggesting that our Havanese dogs, who seem so like shih tzu,[e] might belong to the same genetic clade as standard poodles and powerful Portuguese water dogs.[20] Well, let's be honest. "Surprised" is not an accurate characterization of what I felt. Apparently, vestiges of my big dog ego are still very much

---

[d] That's a period of time between 12,000 and 120,000 years ago, in case you don't feel like looking it up. That's what I'm here for. I like to make myself useful.

[e] The shih tzu, by contrast, may have more in common with Pekes, Pomeranians, and pugs.

intact, as my old pride swelled when I discovered Phoebe and Scout could arguably be classified with those gorgeous working dogs.

The snarky part of me has even constructed an anticipatory retort for the next time some smug guy with a Rottweiler makes a derogatory remark as he passes us on a walk. The comments are always the same and laced with overblown middle-school machismo: a snickered "scary dogs," or "nice fluff balls," or something equally patronizing. My comeback will allow me to stop and pat his dog's sweet lug head while responding, "Sure, Rotties are great. But I prefer working dogs like mine—they're related to Portuguese water dogs, you know."

My fictitious big dog guy will look at me as though I've lost my mind, because he'll never believe that my twelve-pound pups share as much DNA with working dogs as does his formidable companion. And yet, they might. At least according to one recent investigation, the proof of which I keep under my pillow.

So, what does it mean if some small dogs have more in common with some big dogs than they do with other small dogs? What distinguishes small dogs as "small"?

To answer that question, let's first zoom out and consider the miraculous variety in dogdom. Our beloved canines boast vast size differences among breeds. It's a variance that far surpasses that of every other mammal, so much so that research teams have commented on dog breeds' "striking 40-fold range of body size...."[21] But until recent history, no one understood the mechanism of that differential.

Then, in 2007, a group of cooperating researchers discovered the gene sequence likely responsible for small dogs being small.[22] Additional studies confirmed that *all* dogs under twenty pounds share the same genetic instructions that limit their growth.[23] In 2010, scientists found that it's just a tiny number of genes responsible for size and other breed differences.[24] In other words, all dogs share nearly identical DNA, and the genetic material that differentiates them by breed is comparatively minor. To me, what this all means is that very little of what makes up a small dog's essence has anything to do with stature. To butcher Shakespeare's *Romeo and Juliet*, a dog of any size is still a dog. And smells like one too.

New research from scientists at MIT and Harvard's Broad Institute and the University of Massachusetts' Chan Medical School lends authority to my admittedly unscientific theories. While the research team's initial question focused on the connection between breeds and behavior,[f] their paper also reported on aspects of dog size that prompted me to reach out for more information. I wondered about the study's tangential points that, to me, confirmed size has almost nothing to do with a dog's essential nature as "dog." And so, I contacted the study's corresponding authors, Dr. Elinor Karlsson, director of the Vertebrate Genomics Group at Broad Institute and Kathleen Morrill, doctoral candidate at Chan School of Medicine.

Before I relate what I found, I'd like to provide some background on these powerhouse women forging fresh paths in genetics research. Morrill is a scientist in genomics, bioinformatics, and evolutionary biology whose

---

[f] We talk about this more in Chapter Twelve: A Brief Bit of Advice About Finding Your Small Dog.

doctoral work focuses on how genetics shapes canine behavior.[g]

And Dr. Karlsson? She must not sleep, as she not only oversees work at one of the top biomedical organizations in the world, but she serves as a professor of Bioinformatics and Integrative Biology at the University of Massachusetts Medical School. In her spare time, she founded Darwin's Dogs, a massive undertaking that seeks to gather extensive data from dog owners worldwide and to crunch that data in ways that advance our understanding of canine evolution, behavior, and health.[h]

I connected with them both via email, and Morrill provided some additional insight into their research, which, you'll recall, focused on connections we humans attempt to draw between individual breeds and stereotyped behaviors. Their findings, again, exploded centuries of assumptions about how dog breeds behave: Contrary to popular belief, a dog's genetics dictate how she looks and, for the most part, *not* how she *acts*.

And here's where it becomes very interesting in terms of our conversation about small dogs. Initially, when thinking about the connections between small dogs and their behavior, the team hypothesized they likely *would* see small dogs acting differently than large breeds, not because they'd see genetic differences, but because "size might

---

[g] She's also a talented illustrator, and I'm (not so) secretly hoping she might consider illustrating that Young Adult novel that's been making noise in the back of my head for the past year or two. It's about a middle-school girl with a small dog who wants to be an animal scientist....the girl, not the dog.

[h] Not to go all fangirl here, but frankly, I suspect she's in charge of some super-secret time warp lab as well, because, otherwise, I'm not sure how one human could accomplish so much. Either that, or there's Einstein DNA coursing through her veins.

affect a dog's life experiences, and have some indirect effects on behavior (e.g., that the larger world around them might be more intimidating), or that certain behaviors might be a direct function of physical size."

But that's not the case. Morrill and the rest of Dr. Karlsson's team were surprised that "when accounting for breed, the effects of size on behavior mostly disappeared, and any effects of size on behavior were small (<1%)". Yes! What was I saying before? Very little of what makes up a small dog's essence has anything to do with "small."

The team (as most crack scientific teams are wont to do) pushed deeper, as they knew that their findings about size and behavior were muddied by the fact that size and breed are "almost hopelessly confounded." Because a purebred dog's size is dictated by breed, it's not clear which factor (breed or size) is at play when you're unraveling data about a particular set of canines. Accordingly, they focused their attention on their data regarding mixed-breed dogs, where they could sift size from breed.

The result? Morrill and the team "found genetic associations for behavior that could not be attributed to the genetics that affect body size," which "suggests that a dog's size, which is largely shaped by just a handful of genes, is genetically distinct from their behavioral traits."

Dr. Karlsson weighed in, using language that echoes the very heart of this book: "[W]hat we found was that dogs are dogs, no matter their size." She continued, emphasizing "[t]his is a question we looked at very closely. Our genetic analysis showed that the dramatic variation in dog size we see today is shaped primarily by variation in a small number of genes (it's a relatively 'simple' genetic trait), but behavioral traits in dogs (just like in people) are 'complex' genetic traits shaped by many, many genes and a dog's environment."

Small dogs are dogs.

Another compelling factor in my mission to grant small dogs more respect is the fact that, although our small dogs are as much wolf as every other dog, their very smallness can be traced back to ancient wolf lines. One of the primary gene expression sequences responsible for suppressing canine growth existed over fifty thousand years ago. Some experts argue it was in fact the "small" gene coding that came first, and the large genetic material was a mutation.[25] Thus, perhaps big dogs owe something to their elders.

SMALL WORKING DOGS

But enough about the distant past. What can we point to in more recent times to prove that small dogs are dogs? Let's take a trip to the rugged hills of Northern California, where Elaine and her husband Michael[i] lock eyes with the energetic "fluff ball" at their feet.

"You're a farm dog now," they inform their new family member, who peers up at them expectantly.

At four pounds and two months old, Javier certainly doesn't look like a farm dog.

Nor would he grow up to look like one. Not even close.

As a Havanese, he'd stay in the nine-to-fifteen-pound range and stand less than a foot high. No one at the American Kennel Club would ever place Javier in the

---

[i] Names have been changed for privacy's sake.

DOWNWARD SIZING DOG

"Working Breed" category, alongside Dobermans, Great Danes, and mastiffs. Despite Javi's "toy" breed status, he happily adapted to his working dog role, like many other small dogs all over the country. He did, in fact, become the resident farm dog on seven acres among the redwoods in Northern California. And he did so after Elaine and Michael had lived for years with hard-driving working dogs on their farm.

It wasn't originally supposed to be that way.

When their last border collie passed away, Michael informed Elaine there'd be "no more dogs."

They were in their seventies.

She'd just been diagnosed with breast cancer. Michael didn't feel they could handle a new dog.

Elaine begged to differ.

And now Javi, their fifteen-pound Havanese, occupies a central place in their home and on their farm. He accompanies Michael everywhere outside, doing essential farm dog jobs like rolling in cow pastures and course-correcting errant livestock.

Javi is not a border collie. He is not officially a "working dog." But he is a small dog who works well, enriching Elaine and Michael's lives and entertaining everyone he meets.

Much like Javi, my own two Havanese are also lower-case working dogs, guarding our family's domain with vigilance. Before Scout came along and Phoebe's only sibling was her big brother Romeo, Phoebe felt duty-bound to patrol the house on her own. Romeo, with his typical laissez-faire golden retriever attitude, was no help whatsoever, unless "helping" includes enthusiastically greeting potential ax murderers at the door. Truly, Romeo assumed the best of everyone.

Phoebe, smartly sensing that no one else would do the job, assumed the role of resident guard dog. Evening seemed to bring heightened awareness, and as soon as night fell, she would stand guard at the large glass doors opening to the backyard. She'd devote a good hour or two to making noises under her breath every time a nocturnal creature or errant leaf dared stir on her kingdom. We patronized her terribly, smiling at her bravado, but truthfully? We all felt so much safer. No one would ever take us by surprise under her watch.

When Romeo passed and Scout joined our family soon after, Phoebe doubled down on her patrols. Even as a puppy, Scout followed her sister's lead, and now they're my dynamic duo. While they loved guarding the suburban home where we raised our family, they're in security guard paradise now. We moved from our lonely, rather isolated two acres to a bustling neighborhood near the beach. Here, neighbors walking their dogs do so with the same tenaciousness as the postal service: "Neither snow nor rain nor heat nor gloom of night stays these couriers from the swift completion of their appointed rounds."[j]

Nearly twenty-four/seven, no matter the cold, heat, snow, rain, or gale force winds, my amazing fellow neighbors enjoy long walks in the community, usually with

---

[j] Interestingly, while these inspirational words are chiseled above the New York City Post Office building on 8th Avenue, they were originally penned by Herodotus in his work *The Persian Wars*. See "Postal Service Mission and 'Motto'". Historian. United States Postal Service. 1999. Accessed October 20, 2022. https://about.usps.com/who-we-are/postal-history/mission-motto.pdf

their dogs. Just recently, during a rather hard, steady rain, I watched three different neighbors walk swiftly along, their pups trotting happily ahead. Later, when the weather let up to a mere icy drizzle, entire families, including one with three generations and at least two dogs, were out and about. These wonderful community members not only showcase what it means to embrace one's mental and physical health (New England weather be damned), they provide endless reasons for Phoebe and Scout to stand at attention.

And Phoebe and Scout show everyone far and wide that small dogs are, in fact, dogs. They're so good at their guard duties they've become something of a tourist attraction. They may have even entered the realm of minor celebrity status, if you will, as their guard tower makes them visible to all who walk by. Our New England shingled home stands three stories tall, and there's a wonderful reading nook on the third floor facing the street. In that nook is a sofa nestled against two large windows, and the sofa back with its wide, firm pillows provides the ideal perch for two dogs who believe they are masters of all they survey.

They're in their window during the days when we're not home (and often when we are), and they're such fixtures that, blocks and blocks away from home, strangers on the street often stop us on our walks to ask, "Are these the dogs in the window?" I proudly nod "yes," and I've stopped qualifying my answer with "I hope they weren't barking too much," because the answer is always "no, no, just a few times, but then they stopped." The girls know how this whole guard dog process works — they give their alarm, but it's a short one, and they assume we'll take it from there.

Unlike at our old house, when Phoebe shielded us nightly by standing guard at the deck doors even when Ron was home, at the new house, they must assume my husband has us covered. They happily cuddle up with me while I'm writing long past dark, apparently relying on their dad to save the day in case of trouble.

But when I'm home alone, the dogs let nothing—and I mean NOTH. ING.—pass by our house without sounding an urgent alarm. Lest you think "ah...there it is...those yappy little dogs..." let me stop you right there. The girls were so easily taught. They bark. I thank them. And then I say "quiet."

And they're quiet.

Yes, sometimes they need to be told twice. And that's usually when someone's stopped in front of our house to chat with a neighbor or, more frequently, to gaze up at their third-floor lair. The dogs want to be darn sure no one takes me by surprise.

When I'm not home, they don't seem to bark much at all. I'll receive plenty of movement alerts from our security system, but few noise alerts. And our many neighbors assure me that when I'm away, the girls don't bark long when people pass. However, they provide significant entertainment to those enjoying the scenery, because it's not every day you see twelve-pound guard dogs staring down at you from on high.

What does this behavior say? What drives our nine-inch-tall perimeter protectors to watch over our family with such fierce commitment? And why aren't they just lying around like the cat?[k]

---

[k] We don't actually have a cat. But if we did, I'm positive it would be cleaning itself in a corner somewhere. It certainly wouldn't be on high alert against intruders.

To me, it means my miniature guards are fulfilling the ancient human/canine contract. You know the one—it's that document someone drew up over twelve millennia ago, the one where dogs and people agreed to love, honor, and have each other's backs. It binds all dogs, big and small. Size aside, Phoebe and Scout are wolf and canine through and through.

While small dogs can't be expected to take down bad guys with guns, that's not what most of us need our dogs to do. It seems the smarter route would be deterring the sociopath *before* he breaks into our hearths and homes, and small dogs, whose hearing and sense of smell is just as acute as Lassie's, fill that role beautifully. Even the most hardened small-dog-disdainer will agree that small dogs are vigilant watch dogs.

As my two Havanese prove, these tiny canines will spend hours on guard, sounding low, guttural growls when any noise or movement seems amiss. And they're poised to launch into blaring alarm-mode if the situation requires their family's immediate attention.

Small Dog Heroics

Given their status as eager sentries, it's not surprising that small dogs, even toy breeds, can be just as heroic as their larger canine brethren. More importantly, small dogs often can accomplish what big breeds cannot. From being more alert to danger, to fitting where others cannot go, small dogs are our vital partners and deserve an honored place at the twenty-first century "human's best friend" table. To believe otherwise is to misrepresent the power and purpose of even the tiniest dog.

For proof of small-dog heroics, we need look no further than the stories of hyper-vigilant toy dogs who saved their humans from potentially deadly attacks. There's Zoey, a five-pound long-haired Chihuahua from Colorado who leapt in the path of an oncoming rattler, taking the attack intended for a one-year-old boy.[26] And then there's Psycho (these are the names we give small dogs?), a Chihuahua-poodle mix in Texas, who likewise faced down a rattler to protect her young human.[27] In India, a family's Pomeranian died saving her owner from a cobra attack in the dead of night.[28] And across the world in Arizona, Bear, a Yorkshire terrier, jumped in front of his owners just as they embarked on a hiking trail where a rattler was waiting. The snake bit Bear in the chest, but he received veterinary care in time to save him.[29]

A quick internet search uncovers dozens of similar anecdotes about tiny dogs saving their humans from deadly snake strikes. Perhaps small dogs beat out big dogs in this life-saving category because toy breeds are so much closer to the ground and more likely to detect danger at that level. These tales of small dogs putting their human's safety above their own should shush the small dog haters out there.

Of course, small dogs aren't mere snake specialists. Like their large breed counterparts, they possess an olfactory system far superior to our own and just as effective as a big dog's. That acute sense of smell means they often act as early fire alarms. Sometimes, they are the *only* fire alarm, and stories abound of small dogs saving their families from tragic house fires.

There's Jasper, a "goofy" dachshund, who barked his sleeping family into consciousness, saving his human pack from the fire that would have soon engulfed their home. When the barking didn't work, he jumped on his human dad's head, repeatedly, until everyone roused.[30]

Zippy, a Jack Russell terrier, gave his own life to wake his family from a fire so intense it melted the roof.[31] Jaxson, an eleven-year-old Pug, prevented fire altogether when he noticed and sounded an alarm about sparks coming from an electrical outlet in his family's home.[32] And then there's little dachshund Duke, who warned his family, again and again, something was wrong. It wasn't until his third spate of howling and crying that they discovered fire on the second floor above their bedroom.[33]

I know some doubters may think, "These are all believable small dog hero stories: small dog takes on smaller legless slitherer; little dog smells smoke and barks. Surely, though, toy and other tiny breeds can't compete on a larger scale. They can't do what working and guard dogs do every day, filling military or protection roles."

And, yet, sometimes they do.

Perhaps the unlikeliest of war heroes, Smoky, a four-pound Yorkshire terrier, first served unofficially during World War II with her adoptive owner, Corporal William Wynne. Eventually, Smoky's work qualified her to become an official War Dog when she ran telephone and telegraph wire through culverts in combat areas of the Philippines. Her heroics there cemented her name in history.[34] Bill Wynne tells her story in his book *Yorkie Doodle Dandy: A Memoir*.[35]

Another terrier, a mix named Rags, is additional proof that size doesn't matter. Private James Donovan, an American soldier in Paris during World War I, found and adopted little Rags. Initially, Private Donovan thought Rags was just that—a dirty mound of rags lying in the street. Rags came alive, however, and Donovan brought him back to his division, where this little dog's story turned heroic. Donovan trained Rags to run messages between

Allied forces on the front lines, a dangerous job that not only advanced military strategy but saved countless human lives.[36]

These stories prove that, while small dogs aren't the typical choice, they can and do achieve on a big scale in times of war. Their worth as helpmates—and their status as "dog"—shouldn't be diminished because we humans are sometimes blinded by size.

SMALL DOGS AS MEDICAL DETECTIVES

Perhaps nowhere is the "small dogs are dogs" mantra so nicely highlighted as when we look at the way small breeds can prevail not just in cases of human versus human, but in the eternal battle of human versus nature.

Arguably the most heroic protection small dogs provide us lies in their ability to aid us as we battle our biggest natural enemy: disease. In the past decade, research has confirmed again and again that the superior canine nose can detect some illnesses and conditions far earlier and more accurately than any man-made diagnostic. By some estimations, our dogs' snouts are said to be up to one hundred thousand times more acute than our own, detecting substances in parts per *trillion*.[37] Experts equate that level of olfactory talent to being able to sniff out a teaspoon of sugar in water equivalent to that in two Olympic-sized pools.[38]

Such staggering sniffing ability means that dogs might not only help us uncover human disease but also to understand and create more sensitive or accurate medical tests for such enemies. Thus, one company recently trained beagles to find cancers in blood, at an impressive ninety-seven percent accuracy. Results like these will help researchers understand the precise substances dogs are detecting, leading to better diagnostic screening.[39]

Another tragic disease, Parkinson's, currently has no accurate diagnostic tests. However, for years, some Parkinson's medical experts claimed they could always detect a faint Parkinson's "smell" in their patients. Curious researchers sought to determine whether dogs might pick up the scent. Enter small dog Pomeranian Shugga, one of a handful of canines trained to do what our human medicine cannot: determine the presence of Parkinson's.[40] Just like her big-dog comrades, including a standard poodle, a border collie, and the obligatory black Lab, Shugga can detect Parkinson's through scent.

Her small size is no barrier to her service.

Small dogs may also contribute to curbing infectious pandemics, as researchers across the globe work to train canines to sniff out, for example, the presence of COVID-19. They're hoping they can use dogs in public places for rapid detection of infection in larger crowds. The NBA's Miami Heat started one such program, and Florida International University used a team of four scent-trained dogs to detect the virus on surfaces. Two of those dogs were small mixed-breed rescues who can more easily move under and around furniture.[41] So, just as heroics occasionally needs brawn, sometimes it requires pint-sized finesse.

Small Dogs Outdoors

While we're discussing sports, let's shift into one of the most important reasons people crave canine companionship: our dogs' enthusiastic partnership in all our outdoor adventures. Indeed, it's probably the biggest reason people shy away from small breeds or rescues—they don't believe small dogs are actual dogs. They don't believe they can keep up.

And in that, they are wrong.

Let's meet a small dog hiker and sometime marathon runner with (arguably) the shortest legs in the dog kingdom. But before we get there, I'd like to introduce you to Jessica Williams, a dachshund expert whose work with the breed is legendary. Like the funny dachshunds she loves, Jessica's dry wit is clear in much of what she does, including running a wildly popular and irreverently-named website *You Did What With Your Weiner?*[42]

Jessica grew up in the seventies with a forward-thinking, resourceful mom who created a sustainable urban homestead before such oases were a thing. They built raised beds, grew heirloom vegetables from seed, and rescued, raised and/or showed countless rabbits, chickens, geese, cats, and dogs. Jessica's close connection to animals as a child is enviable, though unlike many of the small–dog people we meet in these pages, Jessica chose a one-eyed cat named Jack as her childhood best friend. The family had plenty of dogs as well, but she grew into young adulthood identifying as a "cat person" primarily because she just wasn't in a position to care for a dog.

However, when Jessica entered Western Washington University's Geology program, she developed dog-envy as her fellow students and professors brought their canine companions to the many geological sites they toured. Suddenly, she desperately wanted a dog. But her college-dorm life meant that wasn't going to happen any time soon. So, she turned to volunteering at the local animal shelter to get her dog fix, where she'd do shelter days, transporting adoptable dogs to local pet stores and working with potential adopters to show off the available pups.

DOWNWARD SIZING DOG

In her spare time, Jessica spent hours researching dogs that might eventually fit her lifestyle, which involves everything outdoors and as much rugged hiking and camping as time allows. So, by the time she earned her geology degree, she'd also curated a careful list of three possible breeds that could keep up with her: the Pharaoh hound, the vizsla, and the Rhodesian ridgeback—all dogs that weigh between forty-five and seventy-five pounds.

After graduation, as she applied for jobs, Jessica moved in with a good friend, "Jill". Jessica's stay as Jill's house guest was to be a temporary one, as she bided her time and weighed work offers.

And then, Jill brought home a dachshund puppy.

This is the spot in the narrative where you're thinking, "Oh, here comes the predictable fateful turn, where Jessica drops her big dog crush and realizes that dachshunds are the perfect fit for her."

But if that is what you're thinking, you'd be wrong.

It took a bit longer for Jessica to realize that dachshunds are her heart dogs. About four years longer, to be exact.

In the beginning, Jessica was Jill's built-in dog sitter, as Jill was traveling frequently, for weeks at a time. As a temporary houseguest, Jessica wasn't paying rent, and it seemed the least she could do to help a friend out. But then, Jill's career turned more demanding, and her two-week trips became more frequent. And Jessica's stay turned long-term. Jill was gone on business more than she was home for a good part of over three years. Jessica became Chester's live-in second parent.

Eventually, Jessica moved into her own place, but she still took care of Chester more often than not. And then, "more often than not" morphed into "almost always," until they realized Jill's rigorous schedule would never let up. Jill agreed to let Jessica keep Chester, since the two had become so attached to one another. When the adoption became permanent, Jessica remembers looking at Chester and telling him, "Well, I guess you're my dog now. I suppose I'd better learn more about you."

That was an understatement.

Jessica shares my borderline-obsessive joy over researching, and when the universe (and Jill) made it clear that Chester was all hers, she launched a full-on campaign to discover everything she could about dachshunds. When I asked her why she'd waited four years to do that, she confided that, at some level, she was probably protecting herself a bit. In the years before, Chester wasn't all hers, and some part of her must have believed it wasn't her place to fall in love with his breed.

But then she did. To be sure, before Jessica started educating the world, dachshunds might be the last dogs you'd think of as hiking, camping, and marathon buddies. Even among small dogs, their legs are remarkably stubby. Yet Jessica's life revolves around not merely being outdoors but immersing herself in it. And so, when Chester became hers, Jessica immersed him in that life as well. At the time, she was training for a marathon, and Chester easily ran seven miles alongside her.

When I expressed a bit of shock that a dachshund could keep up such a pace, Jessica shared that while she didn't do everything right in those early years, the running itself is not bad for dachshunds. She knows a woman who routinely runs 5Ks with her three dachshunds, an ultrarunner who's run marathon distances with hers, and a man that is part of her Adventureweiner Dachshund Club who runs organized marathon races with his.

And they thrive.

Dachshunds, while small, are hardy dogs who love the outdoors. Jessica's dachshunds are proof: They happily hike and camp near their Pacific Northwest home all the time. Like almost anything physical for pups or people, it's all in being scrupulously smart about the gear and the gradual training.

We'll hear more from Jessica and her dachshunds in the next chapter on small dog smarts, but for now, let's sum up what we've found on the topic of small dogs being dogs. The evidence is clear. Whether in the role of outdoor adventure expert, guard dog, snake-slayer, firefighter, or medical miracle-maker, dogs like Phoebe and Scout, Jasper, Rags, Shugga, Chester, and a host of other unsung heroes-in-miniature prove it: Small stature takes nothing away from a dog's ability to be our best friend. And in fact, often that smallness is the precise quality that makes them an ideal canine companion.

Small dogs are dogs.

CHAPTER FOUR

SMALL DOGS ARE SMART

Just as many of us mistakenly believe small dogs can't perform the essential dog functions of protecting their humans or accompanying us on our adventures, we also unfairly discredit small dog smarts. And to understand the depth of that unfairness, we need some background.

So, humor me for a moment as I walk you through my once misguided view of human intelligence. I share it because it's possible that you grew up with the same misconceptions regarding IQ as I did. Such misplaced beliefs distort our understanding of other beings around us, including our dogs.

Over four decades ago, as a high school, college, and then law school student, I relished the challenge of standardized testing and embraced the ACT, LSAT, and multi-state bar exams. To me, they were a beautiful game. My brain just worked that way. I loved outsmarting tricky questions and triumphing over bubble sheets.

Later, as a lawyer, I adored wrestling complex tax laws to the ground, and I entered legal disputes with glee. I measured my own mental agility (and that of others) against the standards industrial America had set for me. In short, I grew up with a narrow view of what makes up "intelligence." Not surprisingly, I once applied that concept of intelligence to my dogs as well. To me, intelligence meant powerful executive functioning—things like working memory, impulse control, adaptable thinking, etc.

But then, about twenty-three years ago, I became an English teacher. More accurately, I became a teacher of teenagers. That, and I raised three of my own. And so, I learned that "intelligence" is a funny, elusive thing. I learned that with human intelligence, no one measure exists. Some of the most brilliant minds I know can't ace the SAT or grasp geometry. But they offer us so much more: witty, award-winning writing, show-stopping speeches, block-chain breakthroughs, beautiful art.

The same concepts hold true with our dogs. While certain breeds may be generally more obedient and better able to control their impulses, that fact doesn't give them the corner on smarts. Just as with humans, a dog's intelligence and success in life takes many paths. In this chapter, we'll look at actual measures of small dog intelligence, and we'll see that small dogs match big dogs in the intelligence measures that matter.

But for now, be forewarned that small dogs aren't merely smart. They are sometimes too smart for our own good. In any event, we need to shift our understanding of canine intelligence. Too many of us believe brain size equates to intelligence and that, therefore, large breeds are smarter.

Take Google, for example.

Small Dogs Are Smart

Just try typing "are small dogs smart…" into its search bar, and Google behaves like that obnoxious guy at work finishing your sentences for you. You begin entering "are small dogs—" and before you can get to the word "*smart*," Google jumps in prematurely, suggesting: "Are small dogs dumber because they have smaller brains?"

Well, no.

No, Google, they're not.

Before we launch our full-on counter to Google's misfiring assumptions,[1] let's look at the history of our growing understanding of canine intelligence. Aside from pockets of people throughout time who recognized a dog's essential dignity as a conscious being, most of humanity (and especially "Science," with a capital S) dismissed the idea that canines had true *minds.* Not only did science largely label dogs as mere beasts fit only to experiment on, but they also scoffed at any researcher who dared suggest otherwise.

Now, however, dogs are finally having their day. We're reaching a new understanding of animal intelligence. This evolved viewpoint has birthed a new scientific study of dog cognition. Such focused attention on canine intelligence shows us much about a dog's brain, and it confirms what I've known since bringing my first small dog into the family: Small dogs are as smart as (and probably smarter than) you need them to be.

But before we arrive where we are now, let's visit past eras to understand how far we've come in our

---

[1] Yes, yes, I know Google isn't a real person and that those assumptions are built from Google's knowledge of human search history.

understanding of a dog's mind, and in particular, the power of a small dog's brain.

Early Human Perceptions of a Dog's Mind

Please note: the following paragraphs contain some brief but disturbing references to scientific experimentation on dogs. It was hard writing about it, and I'm sure it will be hard to read. While I've erred on the side of avoiding graphic descriptions, even the brief references to suffering animals might be difficult for some people to read. If you'd prefer not to have this information in your head, skip ahead to the Victorian subsection.

In some ancient, and dare I say, perhaps more advanced civilizations than ours, dogs were often associated with gods of healing or the arts. They were honored family members, given the same care and attention as humans, and sometimes granted full burial ceremonies.[43] They were valued for their companionship and connections to their families.

In horrific contrast, in other thankfully long-dead societies, dogs were deemed mere subconscious creatures who could and should be dissected alive and conscious without concern—all in the pursuit of human knowledge.[44]

And then we "advanced."

Or at least some of us did. For a short time from the third century BCE through the fourth century—but amid continued animal experimentation—thinking people labeled such experimentation "cruel" and "useless." Reformers argued that animal studies conducted while an animal was in pain or dying "would distort the normal appearance of internal organs."[45]

One would think.

And yet still, humanity's self-centered, barbaric behavior toward dogs persisted.

The Scientific Renaissance View of a Dog's Mind

Even when we fast forward to the heart of the European Renaissance, a time of dazzling human creation in nearly every area, we find dogs being tortured in the name of science. It is here we meet René Descartes of the "I think, therefore I am" fame. In my mind, Descartes' thinking was so twisted that we'd all have been better off had *he* never been[m]. Or, at least, any animal within his reach would have avoided unspeakable suffering.

Descartes claimed animals have no soul and were mere mechanical devices with no capacity for feeling. His "animal machine" theory glorified the idea of vivisection, which is dissection while the victim is alive and conscious.[46] He used this premise to justify heinous atrocities on dogs. Perhaps most tragically, he's said to have nailed his wife's pet dog to a board and dissected the living creature without sedation or pain management. The pup's resulting screams of agony were, he explained, merely noises a malfunctioning machine would make.

---

[m] Defenders of this long-ago scientific experimentation on dogs will argue we'd still be in the figurative dark ages had Descartes and his ilk not forged ahead in the name of science. They also defend Descartes' thinking as being "of its time." Yet would we not have made the same advances studying cadavers? And certainly, even before anesthesia was discovered in the 1800s, painkillers could have been employed.

Descartes was not alone in his merciless conjectures. For much of human history, dogs—indeed, most animals—have been treated as having little intelligence or standing, especially within the scientific community. Of course, vigorous pockets of dissent existed. The nineteenth century gave rise to vehement anti-vivisectionists who wrote persuasively regarding the boundless capacity of animals to think and feel; Rene Descartes' niece was one of the most prolific and successful. The scientific establishment largely dismissed these appeals.[47]

There were, of course, exceptions to such dismissals, at least among philosophers. Nearly a century after Descartes' death, French writer and philosopher Voltaire condemned the illogical cruelty of men like Descartes, writing:

> Is it because I speak to you, that you judge that I have feeling, memory, ideas? Well, I do not speak to you; you see me going home looking disconsolate, seeking a paper anxiously, opening the desk where I remember having shut it, finding it, reading it joyfully. You judge that I have experienced the feeling of distress and that of pleasure, that I have memory and understanding.
>
> Bring the same judgment to bear on this dog which has lost its master, which has sought him on every road with sorrowful cries, which enters the house agitated, uneasy, which goes down the stairs, up the stairs, from room to room, which at last finds in his study the master it loves, and which shows him its joy by its cries of delight, by its leaps, by its caresses.
>
> Barbarians seize this dog, which in friendship surpasses man so prodigiously; they nail it on a

Small Dogs Are Smart

table, and they dissect it alive in order to show the mesenteric veins. You discover in it all the same organs of feeling that are in yourself. Answer me, machinist, has nature arranged all the means of feeling in this animal, so that it may not feel? has it nerves in order to be impassible? Do not suppose this impertinent contradiction in nature.[48]

At least some fragments of humanity got it.[n]

Victorian Evolution: Darwin

Moving forward into the Victorian era, we see Charles Darwin's painstaking observations revealing undeniable similarities in the ways animals and humans think and feel. His research gave birth to a revolution in animal study. As early as 1882, Darwin's fellow naturalist and friend George Romanes wrote *Animal Intelligence,* a work centered on Darwin's observational method. The book often ascribed human levels of emotion to animal behaviors.[49]

Romanes' peers roundly criticized his theories.[50] And then someone else criticized the criticizer. And then someone else found fault with the criticizer's criticizer, and so on. Such back-and-forth between and within scientific communities regarding the capacity of animals to think and feel raged on for another century.

Indeed, for hundreds of years, the arguments within the scientific community regarding the merits of studying dogs

---

[n] While I naively thought this whole history review had a happy ending, and I'd be reporting here that our advanced society has moved beyond using animals in experimentation, in the United States we still permit—even require!—that dogs be used to test pesticides, drugs, and certain chemicals. The topic requires a separate tome, so we'll move on.

(and the methods for doing so) revolved almost entirely around what dogs' *bodies* could tell us about animal and human anatomy or medicine. And for most of those centuries, mainstream science rejected the idea of a dog possessing an actual *mind*. Even when science finally acknowledged that dogs were creatures worthy of regard, no respectable researcher could study a dog's mind and remain respected.

The Dawn of Dog Cognition in the Twentieth Century

It wasn't until we neared the twentieth century's end that research scientists could pursue studying "dog intelligence" with any level of support or respect from their peers. The human likely most responsible for sparking this seismic shift in attitude is Dr. Stanley Coren, a Stanford-educated PhD. His brilliant research and prolific writing in the field of human psychology landed him entry as a Fellow into the Royal Society of Canada. It's a rare and high honor reserved for those who've made "remarkable contributions" to their fields.[51]

While Dr. Coren's exploration of the *human* brain earned him widespread professional regard and advanced our understanding of our own minds, his lifelong personal interest in the animal-human bond never wavered. In a 2017 interview with the American Psychological Association, Dr. Coren shared that since his college days, he'd wanted to study the human-dog connection. Unfortunately, as Coren quipped, "Anybody who claimed they wanted to study the human-animal bond at that time was looked at as if they had just gotten out of a flying saucer… There was no way for funding for that sort of thing."[52]

But thanks to Dr. Coren's devotion to dogs, and aided by his distinction in the scientific world, we now live in a world where dog cognition research is commonplace at top universities throughout the world. Besides expanding serious study of dog intelligence, Dr. Coren's "Canine Corner," his longtime column in *Psychology Today,* keeps the topic front and center in the minds of those who study minds.

Dr. Coren is perhaps most well-known in dog communities (or at least most quoted) for his conclusion that most dogs possess cognitive abilities equivalent to that of a two-year-old human—a supposition that led him to publish *The Intelligence of Dogs* in the 1990s. At the time, his professional community warned Dr. Coren he'd lose all credibility if he published a book claiming dogs were as smart as toddlers.[53] But that didn't stop him, and while it might have lost him some professional gravitas, it gained him love and respect from the masses.

And I count myself among those masses. Although three decades have passed, I remember when *The Intelligence of Dogs* hit local bookstore shelves. I recall sitting in a comfy chair at a local shop, cradling that hardcover book with cream pages, feeling so gratified that some brilliant human was providing objective evidence of what most of us dog lovers anecdotally already knew: Dogs are smart beings.

From that point on, Dr. Coren continued to tumble the barriers blocking our understanding of dog cognition. He ushered in a new era in the way we think about how well dogs, well, *think*.

The Twenty-First Century: Every Dog DOES Have Their Day

Thanks to Dr. Coren and many others, the study of the dog mind is now not merely an acceptable area of scientific inquiry.

It's a well-regarded field uncovering mysteries that concern us all.

We now have "canine cognition" centers all over the world, including academic research centers at Barnard,[54] Yale,[55] Arizona State University,[56] Duke,[57] University of Nebraska-Lincoln,[58] Texas Tech,[59] University of Arizona,[60] and University of Auckland,[61] and I could go on, but I think those are sufficient examples for our purposes.

Just researching this brief chapter buried me in scientific articles on dog intelligence, whereas twenty years ago, I'd have come up fairly empty. But today, we have world-renowned animal behaviorists declaring that "[c]anine science is rapidly maturing into an interdisciplinary and highly impactful field with great potential…."[62] These modern researchers argue that studying the dog brain will give us insight not merely into our best friends but into ourselves.

And Now the Main Event: The Science of Small Dog Smarts

Of course, our concern here isn't with dog brains in general. We want to know how smart our *small* dogs are. And arguably the biggest negative stereotype surrounding small dogs is the belief that they just aren't as intelligent as big dogs. But let's take a hard look at the actual science.

In a targeted study of whether large dogs are smarter than small dogs, Megan Broadway answered the question clearly: absolutely not. Her well-controlled testing of spatial memory in small dogs vs. big dogs found small dogs performed just as well as large dogs.[63] Broadway suggests that our misplaced perception regarding large dogs' intelligence arises not because size determines intelligence but because of the tested dogs' prior training and/or comfort with humans.

It's the training concept that always strikes me as being the likely culprit for people believing big dogs have better brains. The truth is, big dogs can cause more damage, so we're much more diligent in training them. After all, no one wants a dog destroying their home. In contrast, we allow small dogs to get away with more, we train them less, and thus they often test as less intelligent.

TYPES OF INTELLIGENCE

Of course, many studies purport to prove large dogs have better self-control and short-term memory than small dogs. For instance, a 2019 study out of the University of Arizona used data from over seven thousand dogs and seventy-four different breeds to conclude that canines with bigger brains have better executive function.[64] However, such studies often ignore one major flaw in their experiments: that what they're testing—a dog's strict obedience or ability to sniff out a treat, for instance—are inbred traits of many working and hunting dogs.

In this rigged system, is it possible they've unfairly skewed the odds in the game of big dog vs. small dog? It's like that saying often attributed (inaccurately, apparently) to Einstein: "if you judge a fish based on its ability to climb a tree, it will live its whole life believing it is stupid." In most lists of "the most intelligent dogs," nine of the ten listed are big working or hunting breed dogs. But is that because our tests are biased, testing a dog's intelligence using techniques that favor a big dog's strong suits?

Through studies like these, we find large dogs *will* seem more intelligent if we measure intelligence as quivering obedience in leaving that treat on the floor. Even so, I'd match my two Havanese against any big dog on any day. Phoebe and Scout will wait at a distance of thirty feet until I give the okay to retrieve that morsel. But apparently, that's not true of all small dogs.

Again, however, is that because I taught my dogs from the experienced standpoint of a former big dog snob? Certainly, many other small dogs don't have the benefit of an owner accustomed to a seventy-pound dog in the room. Often, the studies looking at questions of size and smarts don't worry too much about accounting for such differences.

Indeed, recent work by Maike Foraita, Tiffany Howell, and Colleen Bennett suggests that a dog's early experiences and how we raise them play a big factor in their performance on tests of executive function ("EF").[65] More importantly, the booming exploration of dog cognition often employs methods that muddy the waters, in that such tests frequently measure something in addition to, or other than, actual executive function.

Foraita and her colleagues considered the "laborious" problems inherent in testing a dog's brain. After reviewing research in the field, they noted that bias, small sample sizes, and lack of testing reliability were significant issues. Regarding reliability, these researchers observed the lack of background information collected about the dogs being studied and tests that were so broad they didn't necessarily measure what they said they did. In much the same way that Dr. Coren was inspired to adapt human infant development tests to measure dog intelligence, Foraita and her team hypothesized we could use human executive function surveys as a model for determining canine executive function. And so they did.

Their published work did not differentiate dogs by size, but it did find that dogs who are bred to work with humans, who undergo intensive training, and who have positive early lives score higher in their executive function. I caught up via email with Foraita, who's finishing her doctoral work at Latrobe University in Melbourne, Australia, and chatted with her about small dogs and intelligence tests. Like me, she said she's wondered whether "small dogs' lack of socialization and training is the reason for a lot of behavioral issues associated with small dogs. Lack of training might well lead to less developed EF in small dogs, as we have seen that training influences EF in dogs."

Foraita cautioned me that because she's not done focused research on dog size and executive function, what she says comes purely from her own personal, as opposed to professional, opinion. But given that opinion is shaped by her deep work in the animal cognition field, I'll take it.

Here, Foraita reminded me that, in studying what goes on in anyone's brain, motivation can muddy things. She says "We can try to devise the best thought out experiments to assess individual's cognitive capability. However, if the individual does not succeed in the task, it is very tricky to find out whether it did not have the cognitive capabilities to succeed, or whether it just did not have the motivation to." In other words, she continues, "When dogs don't listen, it is hard to say whether they are not capable, or whether they have simply no interest."

As I delve into in later chapters, there's a stark divide between a dog *understanding* what you want and a dog actually *doing* what you want. As Foraita notes, it's the dog's *understanding* that correlates to intelligence. It's not the performance of the actual task. And too many of our so-called intelligence tests look only at that. Thus, I'd argue that to a great extent, the working intelligence of our dogs depends upon us: Do we provide the socialization, the training, and the motivation to help them succeed, and be happy, in our world?

When we measure *all* types of intelligence (think social, inferential, and spatial), the *size* of the dog doesn't matter. In their peer-reviewed article "Does Size Really Matter? Investigating Cognitive Differences in Spatial Memory Ability Based on Size in Domestic Dogs," Megan Broadway and other researchers found that large and small breed dogs performed the same in many tests of executive function.[66] Here, researchers carefully developed spatial intelligence tests to accommodate size differences. In the thirteen different tests, large dogs and small dogs performed with remarkable similarity.

In addition to looking at how we administer dog intelligence tests, scientists are also delving into the question of brain size itself. They're asking if our longtime societal assumptions equating small dogs with small brains and lower intelligence are accurate. And you guessed it: What they're discovering is that we have been wrong. For instance, one study found that dog brain size rarely correlates with dog body size. Here researchers pointed to the fact that a dachshund's brain takes up nearly all the skull space available, whereas a golden retriever has large sinus cavities inside the "dark cathedrals of their skulls," as poet Thomas Lux would put it.

These researchers concluded that it's not the canine's size that determines brain size. It's the breed. Many small breed dogs have surprisingly large-sized brains, whereas many large breed dogs have brains that are small in proportion to their bodies.[67]

Other research reveals that looking at brain size in dogs doesn't tell us much of use. In fact, one study by Mark Rishniw and Curtis W. Dewey noted that "small breed dogs had a relative brain size far exceeding their 'expected' brain size, while large and giant breed dogs had relative brain sizes of mental midgets."[68] Thus, the study authors point out that a dog as small as three pounds will, relatively speaking, have a brain much, much larger than one would expect.

Researchers tell us this type of brain-size stability within species doesn't exist elsewhere in the mammal kingdom. According to science, no matter how small or big a dog is, their brains must (and do) keep all the functions that make them "dogs." A dog is a dog is a dog. And a small dog is a dog.

We may have mentioned that a time or two before.

According to Rishniw and Dewey's work, this minor variation in brain capacity implies that dogs, regardless of size, retain "sufficient neuronal complexity for the dog to maintain its 'dogness.' Similarly, because large and giant-breed dogs do not behave or socialize differently from small breed dogs, they do not require additional cognitive functions."[69] In other words, big dogs don't have or need giant brains, and small dogs have brains that work in all the ways that make them true dogs.

So there.

Even the dog cognition experts suggesting that large, working dogs are "smarter" than small dogs admit that size isn't a good indicator of *all* types of intelligence. Small dogs do just as well as large dogs in measures of social intelligence. Size also can't forecast a dog's capacity for inferential reasoning (making logical conclusions about a new situation based on previous experience) or spatial reasoning (understanding the relationship between different objects).[70]

We see these intelligences at work all the time in our own Havanese. Our little redhead Scout, who charms her way into everyone's heart, understands that the special head tilt and unwavering soulful gaze she's perfected will gain her a human playmate for as long as she's interested. She also knows that whether we're home or at a friend's house, or staying at a hotel, smacking a door with her paw will gain her instant access to a potty break. She makes inferences about her past experiences (in these examples with people and doors) and concludes accurately that, although the exact human or door may vary, they are similar enough to elicit her desired outcome.

Phoebe's spatial reasoning shines in her ability to create impromptu tunnels out of blankets, pillows, or under and behind furniture. If it's capable of being burrowed through, she'll make it happen. I'd also argue that Phoebe shows remarkable spatial reasoning in connection with her humans. She differentiates between us in the ways she shows us affection, always giving me a kiss in the exact same spot on my forehead, but never to anyone else. And with Ron, she nestles her head under his chin in a way she reserves only for him. Phoebe's distinction in how she sees us in relation to herself and how she relates to us physically shows impressive spatial and social intelligence.

Scout's superior spatial reasoning shines in connection to her own expectations regarding her proper allotment of attention. For example, she understands that when my fingers are on my laptop keyboard, I am in profound-focus mode. She knows I have little spatial awareness in these moments, as I'm not paying attention to anything happening around me (including her) when I'm deep in a writing zone.

When she thinks I've worked quite long enough, she'll hop up next to me on the sofa and smack my hand with her paw to interrupt my typing. And if I don't stop tapping away, she'll sometimes leap over both me and my computer so she can smack my other hand, with greater insistence. If neither solution proves successful, she'll up her game. She stands on the sofa, leaning her front paws on my shoulders so she can look me in the eye. And that always works.

In these instances, Scout shows complex spatial awareness. She's clearly seeing the relationship between my hand, the keyboard, and my attention. She knows that disrupting the connection between that hand and the keyboard is a major key to regaining my focus. And she knows that if one hand can't be stopped, perhaps the other can. She also understands the link between reaching me at eye level and successfully locking my gaze.

Small dogs are smart.

More Stories of Small Dog Smarts

With respect to the concept of *human* cognition, and as I said previously, I've long known what all parents and teachers know: *there is no one measure of intelligence.* Some people are brilliant communicators, others are introverted mechanical geniuses, and others are prolific and profoundly talented artists. Some of us have no particular gift, but we nevertheless manage to be successful and to make the world a better place.

Having owned no less than seven different breeds of dog in the hound, hunting, and companion groups, I believe the same concepts apply to our dogs, including small ones. With that in mind, let's look at some real-life examples in the agile minds of some pretty amazing small dogs.

PARSON RUSSELL TERRIER

Janine Stover Rose has always loved terriers, but her experience with Parson Russell terriers (PRTs) is relatively recent. She'd long admired the breed, sharing that she loved "their size, their character, and their sweet disposition." In 2017, she finally brought one of these sweet pups home and began what will likely be her lifelong connection to these amazing little dogs. His name was Arlo, and between Janine and her co-owner, Arlo proved himself to be a champion in the show ring. He retired as a Grand Champion, in fact, winning Select Dog at Westminster Kennel Club.

It takes a beautifully bred dog to do that. And a smart one.

Janine now owns and handles her own PRTs in the show ring, and she's secretary of the PRTAA, the Parson Russell Terrier Association of America. Janine's work as club secretary keeps her involved and informed about PRTs, and she confides that, "The more I know them, the more I love them. Their wit, humor, and desire to please seem to have no limits. I even taught one of mine how to dance."

Like Havanese, PRTs exhibit remarkable levels of multiple intelligences. They excel at several performance sports, including agility, barn hunt, earthdog, dock diving, and fast CAT.º They are in tune with their humans and deeply affectionate, and they understand the concept of

---

º It's outside the scope of our chapter here to explore these great activities, but you can find information on AKC.org.

time and place. In fact, Janine's PRTs are so socially adept and well-behaved she's been able to take hers to Europe several times.

Having said all of this, Janine knows PRTs aren't for everyone.

But they are for her.

And they *are* for anyone else who adores a trainable, *highly active* small dog. According to Janine and others who love the breed, PRTs are extremely smart, wildly affectionate, and probably as stubborn as they are clever. We won't go into their loving and willful natures here because this is a chapter on intelligence, but if a PRT might be in your future, we'd urge you to visit the national breed club website[71] to learn more.

In terms of small dog smarts, Janine believes that first and foremost, PRTs are problem solvers. She delights in watching their minds work, and she says they're always thinking, especially when they're on the hunt. She recalls witnessing one particular instance of Parson Russell pursuit at their home in northern New England. Janine lives in Vermont in an old house in the middle of the woods. Like most old houses, and certainly most old houses in the woods in winter, hers is occasionally visited by mice looking for shelter from the elements.

One sunny afternoon, a particularly bold little creature had the temerity to poke his pink nose out in Janine's living room as her PRT Daisy slept peacefully on her bed. Janine remembers scoffing at her little pup and muttering that she certainly wasn't very good at her job as a resident rodent patroller, especially since it was full daylight.

But less than an hour later, she owed Daisy a big apology.

Janine watched in awe as her pup not only located this brash mouse but systematically exhausted it, using methodology much like the velociraptors hunting in *Jurassic Park*. Daisy cornered the creature, so there was no way out for the mouse but to make a mad dash for freedom. Unfortunately for the uninvited house mouse, Daisy easily caught it, and with one quick flick of her head, the mouse met its maker.ᴾ And although we might mourn the loss of this little life, certainly Janine's pup was merely doing its job and beautifully: behaving as a thinking hunter.

We might wonder if a dog with whip-smart hunting intelligence could also possess subtler smarts involving an understanding of good conduct in polite society. And I did wonder, aloud, when I interviewed Janine. Yet she assures us PRTs have no such limitations. Her voice rings with genuine pride when she relates their beautiful behavior in the show ring. Despite their wild street smarts, they almost instinctively understand how to be well-mannered in the presence of their breed brothers and judges. PRTs possess a versatility and understanding that there is a time and place for some behaviors that don't belong in other times and places.

What better evidence is there of deep intelligence than a small dog who can regulate behavior based on circumstance?

HAVANESE

Our Havanese are similarly whip-smart, although I wouldn't hire them as hunters. They're more apt to shoo a squirrel or bird out of the yard rather than bother catching

---

ᴾ When Janine related this story, I do admit I was rooting for the mouse. As a child living on a small farm in Michigan, I routinely snuck down to our very scary, serial-killer-worthy basement and sprung the mousetraps my dad set to keep us from being overrun with mice in our equally ancient home. I love mice.

it. I've already talked about their spatial and reasoning abilities, but wait, there's more: Their intelligences also include excellent memories, astoundingly short learning curves, a sharp focus on human instructions, and an uncanny ability to detect people who need their affection.

Phoebe can learn a trick in only one or two sessions, though I admit she's not keen on performing on demand. She's a pay-for-play kind of girl, insisting on high-value treats for her work. In contrast, Scout might take three five-minute sessions to learn a new skill, but once she has it, she'll offer it with glee. And she'll give it for free.

Like the PRTs, our Havanese also possess remarkable social intelligence. They, too, understand time and place with a sensitivity that feels almost human. They just didn't require the level of training our goldens did to comprehend that our surroundings dictate how we behave. For example, they learned in one session that we don't bother wildlife. We can walk past the duck pond—within a foot of a mallard—and Phoebe and Scout saunter by with only mildly curious glances.

On our bayside patio, they'll watch flocks of migrating Brant geese near our shore, and they'll wag their tails in greeting. But they won't bark or whine. They heard me when I taught them "leave it" when it comes to other wild animals. Occasionally, of course, they'll pull on the leash when encountering something new, but a quick, barely verbal "tsch" from me, and they're back at a respectful distance from the wildlife.[9]

Phoebe and Scout's impressive social intelligence likewise shines when we run errands. They trot along,

---

[9] For both the safety of wildlife and the girls, they're always leashed when we might encounter something.

entering pup-friendly stores with an open heart, ready to greet dog lovers, but smartly disregarding humans who ignore them. Because they are keenly observant, they know who is available for affection and who is not.

If we stop for lunch at an outdoor patio, they lay patiently on their rug under the table. And it took very little to train them to do so. They just seem in sync with us, understanding that we're out to eat and it's time for them to chill. We're most proud of their almost perfect behavior, even when another dog goes berserk near them. Phoebe, especially, can become quite disdainful when she encounters what she must consider terribly impolite behavior. Her side-eye at rude barkers rivals the scathing looks the Dowager Countess hurled at impertinent humans in *Downton Abbey*.

Admittedly, they're both wildly hypocritical here, as sometimes on our walks they'll bark inappropriately, deciding that someone wearing a hat doesn't look quite safe enough or that the yellow dog coming toward them must be the ghost of our Romeo. But most of the time, whether they're exercising their brains during lessons, encountering flora and fauna on our daily treks, or out and about in the world, Phoebe and Scout are amazing representatives of the Havanese breed: whip smart, open, and adaptable.

Dachshunds

We met Chester and Jessica of *You Did What With Your Weiner?* fame in our "Small Dogs Are Dogs" chapter, and we're back to learn more of Jessica's beloved dachshunds, who highlight a mental trait that trumps intelligence: unrelenting commitment to one's goal. In fact, perhaps such commitment is the single best indicator of whether our smarts will take us anywhere at all. At least, that's what my math department chair colleague and friend Pat Dulac says: Showing grit and determination—a willingness to look at problems from all angles—is *the* key to success in math and in life.

And, I would argue, in dogdom. And I'll bet Jessica's dachshunds agree.

At least, Chester would have. Jessica relates, with great affection, that Chester had two joys in his life: hiking and food. And not in that order. In fact, Chester possessed a failure-is-not-an-option commitment to food foraging. Not in the wild, but in the back of her Toyota 4Runner. He was so relentless that she installed lockboxes in her SUV to keep him out of the groceries and camp supplies. They *always* used those lockboxes whenever anything remotely edible entered the car until one Christmas when they didn't. Someone had gifted them those Danish butter cookies. You know the kind—they come in a blue tin box whose airtight cover isn't easy to open, and the box is sealed in a thick plastic wrap.

They didn't think for a moment Chester would bother it, and they placed it inside the back of their vehicle without a thought. With Chester and their packages loaded up, they stood nearby enjoying a prolonged goodbye with family. As they chatted, Chester snuck over the back seat, retrieved the entire tin from its place of safety, and proceeded to help himself to Christmas dessert. Well, at least he tried. And would have succeeded if given just a bit more time. When they opened the car door to leave, they found he'd neatly removed the plastic wrap and was millimeters away from successfully having *bent* a cookie-sized opening in the hard tin top.

That's one example of many in Chester's adventures in (literal) dogged determination. Jessica's affectionate laugh is palpable when she describes the way he "would work at stuff and work at stuff" until he achieved his goal. And she emphasizes that such impressive commitment is what makes dachshunds so intelligent. They're problem solvers and very goal driven. Which, she adds dryly, is "great if their goal is your goal too."

Much of the world views dachshunds as untrainable and stubborn. But Jessica, who has worked with or consulted on nearly one thousand individual dachshunds over the years, reminds us it's precisely *because* they are problem-solvers bred to work and think on their own, that dachshunds seem not to listen. Yet in ninety-nine percent of the cases, it's because their owners aren't speaking the right language.

According to Jessica, dachshunds are "crazy scary smart and with the right communication and motivator, they will pick things up right away. They are easy to train but also easy to untrain. If you don't ask them every time…if you don't hold them to it—they learn really quickly that when you ask them to do something, maybe you don't really mean it. And so why should they bother?" Unlike the scientific research that says most dogs match the intelligence of a two-year-old human, Jessica argues that dachshund smarts are more on a par with a teenager's—they both have a way of avoiding behaviors they're not in the mood to give.

I wish we'd known Jessica when we struggled with our own dachshund. I think she'd have helped us find our common language and a means of channeling their grit and independence.

Breed Mixes and Shelter Pups

Of course, purebred dogs do not own the intelligence market. In fact, perhaps small breed rescues who are often an accidental mix of several breeds may benefit from a double dose of specialized intelligence. For instance, my daughter Caroline's dog Rosie, a Chiweenie rescue, combines the sharp intelligence of the Chihuahua with the wit of a dachshund.

Rosie's sense of humor often elicits spontaneous laughter from Caroline and her wife, Alex. And whenever Rosie makes them laugh, even once, she will invariably repeat whatever made the girls chuckle, sometimes days or weeks later. She'll perform the action and then look at them expectantly for the happy laugh. Rosie also learns tricks as quickly as Phoebe—in a matter of a few tries. And if there's a treat to be had anywhere in her realm, Rosie will voluntarily perform every trick she knows in the hopes of a payoff, like some dog slot machine pouring out all she's got. Both this sense of humor and offering of behaviors show a plan of action that demonstrates complex thinking.

Dr. Coren and His Beagle Darby

I introduced Dr. Coren earlier in this chapter, and I was privileged to chat with him in a lengthy email exchange about the dogs in our long lives. As a full disclosure, it's important to note that much of Dr. Coren's past research revealed that, when it comes to intelligence, medium and large dogs (not toy or giant breeds) win. But Dr. Coren is the first to point out that the intelligence he researched is the kind that measures a narrow quality: a dog's working and obedience smarts.[72]

In fact, Dr. Coren quipped he admires the intelligence of retrievers, "and would probably have had a golden retriever or a Labrador except for the fact that I live in a tiny house... and my wife feels that I should leave enough space so that she can live in it too." That certainly seems more than reasonable. And so, as a more than reasonable human, he's kept his dogs fairly small, most recently sharing his space with Ranger, a Nova Scotia Duck Tolling Retriever and Ripley, his Cavalier King Charles spaniel.

While Dr. Coren's life has included many small dog breeds, he says his Cavaliers have been "the most successful." His first one, Wizard, he brought home when the breed was almost unheard of in North America, and the pup was such "a bright obedience competitor and good companion" that Cavies are still part of his world today. Indeed, in one of our last emails, Dr. Coren closed with saying he was wrestling raincoats on Ranger and Ripley as defense against the torrential Vancouver rain that wasn't letting up anytime soon.

In addition to his Cavaliers and toller, Dr. Coren has owned and loved a terrier, who, although small, always believed she was the biggest thing in the room. But perhaps his most surprising choice of canine companion was their beagle Darby, whom he brought home because they wanted a small dog who would be good with their nine grandchildren. The beagle, he says, was a perfect match, given that the breed is "highly sociable, relatively unbreakable, and not large enough to inadvertently knock over a child when it got excited." So why was the beagle a surprising choice?

Well, as Dr. Coren's fellow dog afficionados pointed out, his "own research indicates that [beagles] are seventh from the bottom in terms of working and obedience intelligence." That's way, way down the list of intelligent breeds. His friends gave him "a lot of flak" for that.

We dog people can be so exacting.

Yet, despite what some people would (unfairly) label Darby's deficiencies on the smarts scale, Dr. Coren shares with affection that Darby "worked out wonderfully for the purposes that I chose him, and was a good friend and housemate for my toller, Dancer, and my wife's cat, Loki, as well as being the grandkids' favorite….His only flaw was that he died way too young."

When I asked Dr. Coren a bit more about Darby's life, though, I uncovered more proof that supports what we've been talking about in this chapter: IQ isn't just about obedience. Like all of Dr. Coren's dogs, Darby took part in formal obedience work, with the aim of earning a Companion Dog (CD) title. That recognition proved elusive, though, as he says "On the off-leash exercises [Darby's] nose would touch the ground, find some scent which was interesting and he'd wander off. In those formal obedience exercises, if you give an additional command after the initial one (such as to bring a dog back to heel when he is drifted off) you automatically fail the exercise and thus the whole trial."

But did that stop Dr. Coren? Hardly. With a keen understanding of both Darby's breed and his individual strengths, he brought the little beagle into Rally Obedience, where open and running communication between human and dog are allowed. Here, Darby's true intelligence came alive, and his dad says he "rose up through the rally ranks," ultimately earning his Rally Advanced title. On his way to that badge of honor, Darby earned a perfect score of one hundred.

Dr. Coren shares that "the British Columbia Beagle Club was so dazzled that a beagle got a perfect score in an obedience trial that they gave him a special trophy. The obedience judge who gave him that score was a man by the name of Michelle Calhoun from Québec. He knew about my work and at the end of the trial he came up to me, smiled, and whispered, 'Don't let me hear you bad mouth beagles anymore.'"

In truth, Dr. Coren had never *truly* badmouthed beagles (though it's fun to hear him take "flak" about the issue). As we shared previously, his work on dog intelligence cautions us to remember how many different types of intelligence exist. That beagles landed at the bottom of his IQ test didn't mean he believed beagles are stupid. It means that they aren't necessarily obedient. And ultimately, for Dr. Coren, the most important factor in choosing a dog is their personality. For him (as it should be for most of us), "it is temperament that means the most"; he wants what he calls "a kissy-faced dog" who, regardless of size, "seeks out my companionship."

I couldn't have said it better myself.

Daisy the PRT, Phoebe and Scout the Havanese, Chester the dachshund, Rosie the Chiweenie, Wizard the Cavalier, and Darby the beagle all provide real-life evidence that underscores the new findings in dog cognition: Small dogs are smart. It's all in what and how we measure their particular intelligence and how we nurture those neurons to thrive. Thus, rather than asking Google, "are small dogs smart?", we should ask, "*how* are small dogs smart?" And how can we help that intelligence thrive?

CHAPTER FIVE

SMALL DOGS ARE POWERFUL HEALTH BOOSTERS

Perhaps the place where a small dog's intelligence far exceeds our own is in their seemingly limitless capacity for understanding and connecting with their human families. But that's not to say that small dogs are more empathetic than big dogs. Not at all. To be sure, all my dogs, big and small, became fundamental components of my happiness and well-being.

My Irish setter Meg's constant companionship gave me dignity and a strong sense of self when I suddenly grew five inches taller than my friends and morphed into an awkward, gangly middle schooler. My first golden helped me raise my kids, and my last golden helped me let them go as they fled the nest. All these big dogs sensed when to lean in—when to offer me an extra dose of their ever-present compassion.

Our experience with Phoebe and Scout is the same.

But it is also more.

Perhaps because small dogs so often resemble human toddlers in their frequent requests to be picked up, or perhaps because of small dogs' close physical proximity to us in our chairs and on our beds, they offer a more frequent and sustained comfort than any big dog I've ever loved. Admittedly, my experience is anecdotal. But it's also backed up by the ways small dogs help humans with more than mere companionship. Thus, for example, expert trainers often choose small breeds as medical support dogs. In many instances, their lighter, tinier frames are better suited to supporting their humans.

One key to the healing small dogs provide lies in their endless capacity for spreading laughter. Of course, we agree big dogs are likewise talented. We remember spending vast blocks of time enjoying our goldens and their antics: the hours Romeo would happily lie in the snow with his iced-over, neon-orange tennis ball perched protectively on his paws; the laugh-out-loud sibling rivalry we watched as two enormous dogs vied for the too-small space on the sofa next to us; and the not-so-surreptitious way Romeo would casually rest his head as close to one's snacks as possible. All of that contributed to so much shared family amusement.

But really, something about small dogs is all of that, and so much more.

Small dogs provide a miniaturized, in-real-life *National Geographic*. Ron and I spend an embarrassing number of minutes that add up to hours leaning against one another (and our kitchen island) gazing at Phoebe and Scout wrestling with each other or their toys. It's the same thing we did when the kids were toddlers—we'd watch them engrossed in complex play, taking themselves very seriously, and then we'd look at one another, enjoying the shared knowingness.

One of the profound experiences of parenting *human* children is witnessing the miniature beings we've created engage in play that mimics adult life. We see them wishing to be us, and we know the reality is both greater and lesser than the fantasy. The experience almost stops time. Watching small dogs "be" in the world produces similar moments of suspended time.

Because small dogs often look like perpetual puppies, we're both entertained and moved by their incongruously fierce treatment of errant socks, their rearing up on hind legs to headlock a sibling, or their guttural growls as they spot a dangerous squirrel daring to enter their domain. There's something deeply calming and hopeful about watching a small creature do big creature things. It produces a circle-of-life type lump in your throat. Those daily reminders of life's bigger picture bring us out of our stressors and into the powerful and healing present moment.

Small Dogs Ease Situational Stress

In my small circle of family and friends, I know of at least three perfectly successful and well-balanced humans who likely would be neither successful nor balanced if it weren't for the small dogs in their lives. These people have careers that can only be described as grueling and at times emotionally horrific. They go to their jobs and grapple with humans who are desperately ill or frighteningly unbalanced or uncontrollably angry. Or all the above.

And then they are to leave work, go home, and find ways to restore themselves so they can return the next day and be strong for others.

Yes, they have loving human relationships and people with whom they can unpack the day. But often, the last thing we need is to relive the trauma of addressing someone else's trauma. And certainly, simply unloading our emotional baggage from work onto our human family doesn't serve anyone. Where does that stress and heaviness go?

It seems to dissipate in the presence of our small dogs. They provide a gateway to home that allows us to let go of our work world. Their effusive, never-fail greetings and snuffling kisses make it hard to do anything but experience joy breaking into the day. Their cheerful company as we move from room to room offers quiet companionship. Their forehead-to-forehead mind-melds simply melt away the need to worry over what we cannot fix. They bring our hearts to a slower, steadier beat and our brains to a calmer place.

Of course, it's not groundbreaking to say that dogs improve our mental health. But the precise hows and whys our pups help us emotionally are still a relatively new frontier in the study of human-animal interactions. In the past two decades, however, researchers have begun unlocking the chemical secrets surrounding our bond with our dogs. What they're finding proves why so many of us feel better in the presence of pups.

Current research tells us that cortisol (the so-called "stress hormone") and oxytocin (the "love hormone") are the primary chemicals involved in the human-canine relationship. In fact, as most pet parents could tell us without the benefit of science-stamped research, the emotions humans experience when bonding with their dogs are like the joyful, primal rush mothers experience with their babies.[73]

If we must throw around scientific terminology, however, we can talk about the way our dogs activate our oxytocinergic system.[74] This mouthful of a phrase describes a self-perpetuating emotional loop that creates feelings of happiness, warmth, and connection while simultaneously reducing feelings of anxiety and stress. We experience this feel-good cycle both from our fellow humans and from outside our species.

In fact, when we interact with dogs (especially our own dogs), our oxytocin level shoots up measurably (as does that of our pups).[75] The "oxytocinergic system"—this lovely psychological and physiological phenomenon with the decidedly unlovely name—improves the quality of our everyday lives. It calms us; it infuses us with warmth and feelings of belonging; it makes us happy.

While the physiological effects of interacting with our dogs produce the hormonal results that make us feel better, sometimes those physiological effects are not the crucial thing. What *is* crucial is the way dogs provide acknowledgment and concern over our emotional pain in a way that humans cannot. Especially with situationally specific stress, a dog acts as "a comforting, nonjudgmental presence."[76] Even when the humans we're reaching out to *aren't* being judgmental, we, in our stressed-out states, often perceive or assume such judgment. If we're wondering what the other person is thinking about us, we can't get true relief from a human companion.

Enter our small dogs.

Studies confirm that, when we're depressed or juggling crushing stress, dogs offer us non-judging, selfless, and reliably available "social recognition."[77] Indeed, when it comes to helping us cope with stressful situations, one study found that dogs do a *better* job than our human friends: Dogs can lower our heart rates and cortisol levels more effectively than our human BFFs.[78] Another recent study found similar results in measuring dogs' calming effect on stressed out college students.[79]

SMALL DOGS AID SLEEP

The same thing is true in the bedroom, where sleeping with our dogs can improve our sense of well-being. Especially for women who have issues with sleep, sharing our beds with dogs can improve sleep quality. According to researcher Christy Hoffman and her colleagues, women who allow their dogs to sleep in bed with them report feeling more secure and comfortable; in addition, our sleep patterns improve because we tend to go to bed earlier and rise earlier to match our canine family member's sleep pattern.[80] I can attest to these experiences and am among the small dog owners whose slumber is infinitely better because we now have both our pups sleeping with us at night.

And the pups have improved my sleep habits, although sometimes when I want to live dangerously and stay up past nine o'clock, it's a challenge. Ten-pound Scouty has a bedtime warning system. She tries to stay awake, but her lids will close on her and if she's sitting upright, she'll teeter precariously. I'd swear she'd tip over if I didn't lend a finger to steady her. By nine p.m., she needs to go to bed. And, because I must be up well before six a.m. on weekdays, I'm happy to oblige. In fact, I cherish our bedtime ritual, which gently separates us from the day as our faces are washed, my watch and their collars are set aside, and our bodies are tucked in until morning.

During the night, the girls snuggle along the length of my body, with Phoebe near my hips and Scout along my side. I often stir a bit in the night, reaching out to pat them, an action that soothes me back to immediate sleep. Scout always obliges my disturbance by turning her tummy to just the right angle for a possible skritch. Their presence, like my husband's, is a touchstone in the night. We could never enjoy such family bedtime with our goldens. Ron, his bad back, and two body pillows claim three quarters of our large bed, leaving very little space for me. And, even if there had been room, I couldn't have borne the hair in my bed.

Of course, dogs don't *have* to sleep on the bed. And many big dogs sleep very well in their own bed on the floor. But this book is about the many advantages of small dogs, and one of those plusses is the way they can improve the quality of our sleep by being part of the rhythm of our nights. Big dogs cannot give us that, unless you enjoy giving up significant mattress real estate and don't mind being unable to move the comatose boulder on your covers.

Small Dogs Improve Our Mental and Physical Health

Dogs are emotionally therapeutic for all of us. But dogs can be literal lifesavers for those of us suffering emotional or physical challenges. Whether in our homes, hospitals, rehabilitation facilities, or cancer or stroke treatment centers, pups acting as therapists or service dogs lessen human suffering wherever they go. The silent support dogs provide—their warm touch, their unwavering gaze, their (usually) agendaless empathy—reaches something buried in a human's injured psyche.

With nursing home patients suffering dementia or Alzheimer's, a decade-old but well-controlled Italian study revealed that dog therapy reduced patients' depressive symptoms by *fifty percent*.[81] A much more recent study found that dogs can help older, institutionalized dementia patients feel happier and engage more,[82] and other researchers reported significantly better patient interactions and even mobility when dogs are involved in the therapy.[83]

Similarly, a Swiss study revealed marked improvement in objective measures of cognitive and memory function in Alzheimer's patients who engaged with dogs in animal-assisted therapy sessions. What researchers found striking is that the benefits of those sessions drop off when the dogs stop visiting, and they recommended the dogs should be a "steady presence" in the patient's life to achieve sustained benefits.[84] Why this would surprise them, I don't know, given that most of us only benefit from relationships when the relationships actually continue. But maybe that's just me.

Dogs provide us with similar benefits during physical illnesses. In hospitals and cancer-care centers, it's well-established that patients experience significant emotional support from therapy dogs. Even in the face of brutally painful chemotherapy and radiation treatments and their aftermath, patients visited by emotional assistance dogs experience more well-being. A recent study aimed at ensuring the health and safety of immune-compromised pediatric cancer patients reviewed a long list of benefits visiting therapy dogs provide, including lower blood pressure, better moods, and requests to see the dogs more frequently.[85]

Animal-assisted therapy for all cancer patients is widely recognized as beneficial at every cancer stage, including treatment and hospice. Indeed, the Penn State Cancer Institute has allowed and encouraged therapy dogs since 2005.[86] Sloan-Kettering Cancer Centers began employing animal-assisted therapy in 2007.[87] Mayo Clinic's Caring Canines program started in both their Arizona and Florida locations in 2011.[88]

Given the top docs who invite canines into their hospitals, it's not surprising to learn that therapy dogs simply make life better for patients and their families. And while I sound like a broken record, small dogs can do all this much more conveniently than their retriever counterparts. If they're non-shedding dogs, they can also do so with less allergy risk and less mess. When they're small, they can also make closer contact with patients, providing better emotional connection and comfort.

In our family's experience, that comfort level often has nothing to do with a patient's discomfort with or dislike of bigger dogs and everything to do with the perceived danger a big dog can pose, given the patient's medical condition. When a close friend of ours was battling cancer, her beloved retriever was no longer permitted to lean against her as he once did. It caused her agony when he tried. She was dealing with chemotherapy ports and debilitating nerve pain. Cancer treatments had destroyed her ability to keep much food down, and she was becoming so frail. She confided that even though she knew her loyal, loving dog would never hurt her, she was so fragile that she tensed up every time he was near.

And that made her pain worse.

She didn't have that same fear with our Havanese Phoebe, who was under ten pounds at the time. She loved Phoebe's comfort and warmth, and I know Phoebe eased her pain. I swear, she invited me over all the time, not because she needed my company, but because she needed Phoebe. And that was more than fine.

Our story with Phoebe is an example of a small dog who took to her role as a happenstance therapy dog, and thrived in it, but then returned to being "just" our family member after we lost our friend to cancer.

In contrast, other small dogs make service work their life's mission. The field of Animal Assisted Interventions (AAI) is now so well-established, and the benefits to humans are so clear, that current researchers focus less on whether animals benefit *patients*, but on how we can ensure therapy animals' well-being. Now, we're at the point where the American Veterinary Medical Association (AVMA) has established guidelines for overseeing a therapy dog's welfare, including required annual wellness exams, training for owners, and assurances that the handlers protect their dogs during sessions.[89]

In fact, nearly all human-animal interaction studies I've read from this decade take care to detail how researchers safeguard pups during studies of AAI work, including one study where elementary-aged children took part in special training before the dogs set foot on school grounds.[90] Given the delicate nature of many smaller dogs, it's likely they'll develop additional precautions for these breeds. Having said that, as the Americans with Disabilities Act National Network tells us, *any* size or breed of dog can and does act as a service dog.[91]

A quick Google search will put you pages deep in examples of small breed dogs who work as service animals, providing hearing help, object retrieval, and even bed-making services for their humans with physical challenges. With respect to emotional support, small dogs are especially adept at providing specialized care for people with post traumatic stress disorder (PTSD).[r]

Carla and Riki

That's the situation with a wonderful Havanese we know. Riki is a ten-pound certified service dog whose presence helps my friend Carla construct a life well-lived. We met Riki and his guardian Carla through Instagram. Initially, I reached out to her because I loved her work on Riki's account, and I just sensed a kindred spirit in the pet space. Her aesthetic, her humor, and her positivity on Instagram all drew me in. I hoped that we'd find some common ground as women with a shared love not just of small dogs, but of how small dogs can transform lives.

Carla and Riki live in Utah, about twenty-three hundred miles away from us. But because the pandemic successfully erased most worries about long-distance business relationships, Carla and I scheduled a Zoom exchange. And we just clicked. Our calls were full of strategy sessions and exchanged ideas regarding ways to support small dogs and their people, and we quickly developed an easy give-and-take mix of personal and work stories.

Although I'd learned early in our chats that Carla had undergone open-heart surgery and as a result had given up her high-driving career, it wasn't until our third call that

---

[r] Many mental health advocates and veterans' advocacy organizations assert the more appropriate terminology is "post-traumatic stress injury." Since the accepted mental health term is still PTSD, I've used it here. Nevertheless, I want to recognize the compelling reasoning regarding a change in terms that recognizes the real biological injuries inflicted by trauma.

Carla briefly referenced her PTSD stemming from an assault related to that surgery. I didn't believe it was my place to ask for details, and now that I know a bit more about PTSD, I'm glad I followed my first instinct.

It turns out that often, asking someone to relate the event that gave rise to the PTSD can trigger severe symptoms and what is essentially a re-traumatization experience. The most important part of the story—Carla's recovery path—is one that deserves its own book, which I'm hoping Carla will write. Here, what's important to relate is the PTSD Carla battles in such a heroic way with Riki in her corner. In these pages, we look at the way Riki's service work is vital in helping her do so. But first, some background.

Like most of us who grew up decades ago, Carla's childhood dogs were large family dogs. They had Beau and Duke, two great Labs, along with another large mixed breed they adopted. A few years later, Carla's family adopted Lady, a small dachshund mix, who could tag along to their video rental shop (remember those?) at the base of a Park City, Utah ski slope. While Lady spent her days as a savvy retail dog, she also became Carla's best friend. Despite Carla's significant dog allergy, Lady slept with her and kept her sane through her early adolescence. When Lady disappeared after a trip to work, Carla's devastated thirteen-year-old self swore she'd never love another dog.

And then Carla grew up and had kids of her own. They lived in rural Utah, where everyone had a dog and everyone's dog spent most of its time outdoors. And so, the kids had to have dogs too. Their dogs were McNabs, a herding dog developed in Northern California, and these pups mostly belonged to the kids, as Carla still wasn't really recovered from the loss of Lady.

And then she and her husband divorced, a life event which hit her children with unsurprising force. One afternoon, shortly after the separation, their youngest son arrived home from the park with a boxer puppy in tow. The puppy stayed for three weeks as they looked for the owner, and it devastated her son when the owner came to collect the little guy. Not long after, the universe directed Carla down a street past a home with a sign in the yard reading "boxer puppies."

She stopped. She saw the runt. Minutes later, she was driving away with the puppy in a box, heading to pick up her son from his friend's house. When he hopped in the car and saw the puppy, he burst into ten-year-old-boy tears of pure joy.

Of course, they brought Coda home exactly the way you're *not* supposed to do it. No plan, no forethought…no food. But again, it was in a time and place when dogs ran loose outdoors and ate table scraps (not that table scraps are necessarily a bad thing). Coda became another heart dog for Carla, and it crushed her when her son and the pup went to live with their dad full time.

That was in 2001. During the next seventeen years, Carla's career exploded as she managed major markets in the insurance industry for several Fortune 500 companies. I call those her desert years, because although her work life was full, she wasn't coming home to a dog. It was toward the end of this two-decade period that a devastating trauma led to Carla's PTSD, which manifested as night terrors, anxiety attacks, episodic dreams, and crippling sleep deprivation. Most people know post-traumatic stress disorder as something experienced by soldiers after war, but it's a condition that can develop after any terrifying encounter,[92] whether in combat, after a violent attack or an experience with death.

Despite her diagnosis and amid relentless symptoms, Carla channeled her stop-at-nothing corporate self to aid her own recovery. She researched tirelessly, determined to build a vibrant new life *with* and *despite* the symptoms she battles daily. And her research revealed that building such a post-trauma life requires many levels of help, including direct therapy and complementary healing. For the first time in nearly two decades, Carla turned to help in the form of a four-footed friend: She knew she needed a service dog.

In her investigations, she narrowed her search to non-shedding dogs, to ease the likelihood she'd suffer allergy issues. And, whatever the breed, the pup had to be whip smart *and* meant to be a companion. Finally, she wanted the pup to be small. Knowing how much her childhood dachshund, Lady, comforted her during her early teens, Carla was drawn to another small dog. Her research led her to the Havanese breed, after reading a UK study that found these little pups beat out golden retrievers, Chihuahuas, cocker spaniels, and others when it comes to comforting anxious humans.[93]

Within a year after reading that study, Carla found Riki the Havanese and brought him home. His breeder has since passed away, but Carla chose her in part because every year, the woman donated one of her Havanese puppies to a local military base. These small dogs were also trained as service dogs to help spouses and children of soldiers—and sometimes the soldiers themselves. Carla was also drawn to this breeder because, in addition to Riki's impressive working pedigree, his father is a top show dog in Europe. Carla knew she also wanted Riki to help be an ambassador for small service dogs, so a pup with the confidence and flair of a show line was a bonus.

Normally, you'd wait several years for a dog like Riki, and he was supposed to go to someone else. Unfortunately for them, but fortunately for Carla, the family slated to have Riki could not take him when their turn in line arrived. The breeder called Carla, who was next in the queue. Within days, Carla was off the waiting list and Riki was on his way to Utah.

From puppyhood, Riki began his service dog training. He began with in-home visits from a positive behavior specialist. Carla calls those initial lessons "life changing," as they improved and cemented their way of being with one another. Then at five months old, Riki went to a residential service dog trainer for a two-week stint in which he lived away from home and Carla wasn't part of the work he was doing. In looking back, Carla says she'd likely find a different way to certify Riki as a service dog. It was too much time away, and she says with service work, it's frankly as important to train the handler as it is to train the pup. The stay-away camp didn't provide that.

What it *did* provide Riki, however, was a solid base in service dog skills—expertise that he uses almost daily to help Carla grapple with (and usually triumph over) PTSD symptoms. Among the worst such symptoms are the episodic dreams that have Carla reliving her original trauma. Here, Riki will wake Carla by nosing her neck and bringing her out of the experience. She calls it "the most powerful thing...having this other loving being bringing me gently back into the real world."

Afterward, Riki will lie on Carla's chest, much like a weighted blanket. Except unlike a weighted blanket, Riki can match his breathing to Carla's. Or rather, she will match hers to his. She might be terrified and in a panic attack, and he will lie with her for twenty to thirty minutes, until their breathing syncs in slow, calm cadence.

Before Riki, Carla would suffer in the dream with no one to rouse her, and then when she reached a waking state, she'd often descend into a panic attack, or worse, she'd lie there sleep deprived, unwilling to shut her eyes and fall back into the terror. But with Riki — his firm presence, his knowingness of her stress, and his connection to her—her trauma eases, and he keeps her in a safer place. And Carla says he *knows* he is helping. He is both satisfied in and proud of doing his job.

Riki's happy work continues during the day, where sometimes Carla's sleep deprivation can trigger anxiety episodes. When that happens, she has a process. And Riki is a vital step in that process, so much so that often he senses an attack coming on before she does. Often, she'll absent-mindedly wonder why he's shadowing her so closely, and then she'll realize her breathing is off.

And her process begins.

She paces and works to steady and deepen her breath. If she reaches the point where the breath isn't coming, she'll grab a soft toy to play fetch or tug with Riki, and that will usually bring her out of the episode. If she's still struggling, she'll take medication and go lie down. Riki joins her and just petting his long silky hair calms her immeasurably. Before Riki, these episodes could last four or five hours. Now, they're often subdued in less than an hour.

What is most impressive about Riki and his support work is his intelligent adaptability. He behaves like a canine medical professional, attuned to his human's vital signs. Recently, Carla experienced Riki repeatedly hopping on her lap and snuggling her neck like he does when he wakes her from an episodic dream. She kept asking herself (and Riki) what on earth he was up to—it simply wasn't a normal thing to be doing when she was awake and not in a panic mode. She'd put him down, but he'd follow her, refusing to leave the room and staying within feet of her, checking on her repeatedly.

And yet Riki knew exactly what he was doing.

He clearly sensed something was wrong, and he was right. It took Carla and her so-called smart watch longer to get with the program. It turned out Carla's heart had kicked into overdrive, racing away with a bout of dangerous tachycardia that Carla hadn't sensed but Riki had. This little Havanese had not only diagnosed the problem, but he'd attempted to treat it with the calming techniques he'd already learned to use in other situations. (Talk about a small dog with inferential reasoning powers).

Thus, inside Carla's home, Riki is a hero. Outside her home, he is her lifeline. Carla's trauma caused trust issues that make it very difficult for her to feel safe around strangers. She relates, "I don't want to be seen, I don't want to interact." And of course, that feeling can be problematic at best, preventing Carla from accomplishing basic errands, and crippling at worst, cutting her off from the good parts of humanity.

But then Riki puts on his service dog vest and understands he's now at work in the world with Carla. His attention is on that important job, and he keeps a perimeter around her, placing himself between others and Carla. And she laughs because she says it's hard, but in a good way, to have a cute service dog. Riki becomes the topic of conversation and distracts her from intrusive thoughts and fears about new humans.

While some types of service dogs (police and sight dogs, for example) require at least a year of training before they're ready for their jobs, that's not necessarily the case with small companion service dogs who work with humans to improve quality of life. They often adapt with less regimented training, as Riki shows us. Because of Carla's health, the pandemic forced her to remain isolated, so Riki's training was interrupted and then not consistently used. She and Riki rarely went into public, and for two years, he didn't put on his vest or accompany her around others.

When the pandemic eased a bit in the spring of 2022, Ricki hadn't truly been out and about since he was about a year old. Neither had Carla. But a grandchild's upcoming birthday inspired them both to venture to the bank and post office. Carla admits she had no idea whether Riki would remember any of his training for such outings. But when she put on his service dog vest, it's as though not a day had passed since his prior work.

When they arrived at the bank, Carla blanked on her PIN and was forced to go inside the building. She put on Riki's vest and crossed her fingers that he'd behave. She needn't have worried. Not only did Riki behave, but he also transformed into serious service dog mode, going into an alert down/stay next to her during the half-hour it took in the manager's office resetting her PIN.

Later at the post office, she brought him inside and Riki remembered his perimeter work, always placing himself between Carla and anyone near her, a feat that's complicated when you're standing in line with humans in front and behind you. But Carla recounts how "phenomenal" Riki was, keeping eye contact with her and responding instinctively to her cues. When two older women in line asked if they could pet him, Riki also looked to her for permission before stepping forward.

Despite his nearly two-year hiatus from public service dog work, Riki's training all came back to him.[s] He's a shining example of how small dogs prove to be powerful therapy partners for all of us. Riki is so transformative for Carla that she's named him her "Ability Pup," which is Riki's new Instagram handle and Carla's nonprofit company. In the coming year, they'll be working together to help other support dogs and their people.

Small Dogs at Work

For many of the same reasons dogs aid those of us struggling with physical or mental health issues, they likewise benefit us at work. Just as hospitals and medical facilities use Animal Assisted Intervention ("AAI") for patients in health care settings, organizations are bringing AAI into high-stress environments to benefit front-line workers. A literature review spanning 2001–2021 reveals the beneficial impacts of animal therapy in health care settings—not just for the *patients*, but for the *caregivers*. The study reveals not only are such programs doable, doctors and other medical staff readily accept "the immense psychological benefits [they] provide…".[94]

Of course, formal animal therapy programs for first responders and health care workers are wonderful. But most workplaces don't need such targeted programs to

---

[s] We'd be wrong to give the impression that small service dogs are automatons who always know how to be in the world. Riki did develop one embarrassing foible—some dog reactivity created due to their COVID isolation. But Carla reached out to their trainer again and Riki quickly regained his composure around new dogs.

reap the many benefits of a dog-friendly workplace. Let's look at the thousands of dogs who informally accompany their families to work daily. At Amazon's Seattle headquarters alone, over eight thousand dogs chaperone their people as they go through their workdays.[95]

The company and the employees (and their pups) agree having dogs as part of their workplace culture isn't just about reducing stress. Amazon managers report the surprising ways in which dogs bring people together, noting they provide means of connecting that don't exist in dog-free environments. People who wouldn't otherwise stop to talk find a reason to do so when coming to and from their offices.[96] Small talk occurs easily, and in the best instances, grows into genuine conversation about important work. All because of dogs.

The positive effect of dogs in the office caught the eye of researchers, whose studies confirm the many benefits of pups at work.[97] A recent case study confirms that a business who carefully implements a policy allowing employees to bring their dogs to work will see an incredible positive impact not merely on the single employee, but on the entire work culture.[98] Dogs at work make life better not just for the dog's owner, but for everyone around them and thus for the company itself. Google is a case-in-point, where dogs were first welcomed way back in 1999.

Now, Googlers who regularly bring their pups to work run "Doogler" groups at Google offices worldwide. In a twist on brand pop-up shops, dog-loving Googlers began creating "pup-pop-ups" at work, giving their dogless coworkers time to destress with adorable, wriggling puppies during the day. And during the pandemic, the Doogler groups were key means of maintaining connections during work-from-home.[99]

Other employees at companies like Etsy and Replacements talk about how much easier it is to get to know coworkers through their dogs, and how much laughter and comedy they bring to the office.[100] And these companies are not alone. In my many Havanese contacts through Instagram and Facebook, I've heard countless happy stories of dogs at the office. While I, of course, advocate for allowing *all* well-trained dogs at work, in my mind small dogs have an edge. As we talk about in the next chapter, they're cleaner, easier on allergies, and better for the environment. They also are much less likely to be intimidating to coworkers who don't (yet) love dogs.

Small dogs can charm their way into the heart of even the most reluctant human much more efficiently than almost any big dog can. The fright factor simply isn't there. It's hard to fear someone who stands eight inches high and a toddler must reach down to pet. So, when a small dog approaches us at work, we feel more inclined to interact with them. And boy, our small dogs take advantage of that.

We heard of one Havanese who brings her favorite golf ball to every office visitor, inviting guests to roll the hard white orb down the hall for her to fetch. You'd be hard-pressed to allow such indoor games with a retriever and a tennis ball unless you enjoy out-of-pocket expenses to repair the ensuing destruction. A small dog though? You can throw the ball for them all day, with no damage in sight. The natural icebreaker this pup offers everyone, and the smiles she elicits from visitors, must create a sense of immediate shared experience that builds stable connections both for the present and into the future. A client is less likely to feel dissatisfaction while interacting with a business's furry ambassador. Indeed, a business may well gain more clients as people drop by just to say hi to the little office worker in a small dog suit.

RIVER AND BROOKE

River, another Havanese we're lucky to know about, is such an important employee that his name is displayed on the office door. And on the lease.

Seriously, the lease.

River works for Converge Local, a forward-thinking creative agency based in Virginia. Brooke Anderson is River's owner/guardian, and she also happens to have a side job as chief convergence officer, bringing together all the firm's departments into a cohesive whole to best serve its clients. But before we get the details about River, let's learn a bit more about Brooke, a human we wish we could clone. The world would be a better place. Except for the, you know, whole cloning thing, which never ends well, at least in every sci-fi story ever written.

To understand River's impact on others, it's important first to understand Brooke and her drive to help companion dogs fulfill the jobs they were meant to perform: to be meaningful, helpful human companions. She's drawn to Havanese in particular, because they echo her own commitment to showing up every day expecting the best of every situation.

Brooke credits her mom's example for her desire to help others. During Brooke's childhood, her mom worked in social services, and every Thanksgiving, her mom brought her along to deliver food baskets to needy families. Brooke says that initially, she felt like a generous Santa Claus, rewarding people with riches. But then they reached a home where a boy her age opened the door.

In that instant, her understanding dawned as she recognized his obvious embarrassment. His discomfort taught her the most important part of helping others: avoiding, at all costs, making people feel unworthy. She identified with the embarrassment in the boy's face, and that recognition spurred her realization: We're all within one tragedy of needing such help ourselves.

Since that childhood moment on the steps of someone else's house, volunteer work became integral to Brooke's life. And while she can't pinpoint the reason, she says she always knew having a small therapy dog would aid her in bridging the artificial divide between the human helping and the human needing help. So, when COVID-19 hit and she was working primarily from home, the universe conspired to fulfill this dream of hers, and puppy River joined her home.

During the partial COVID lockdowns, Brooke brought River with her to Converge, where everybody loved him. The feeling was mutual. When they neared the office, River would race to the door to greet everyone, thrilled to spend a day at work. His enthusiasm as the "happiest employee ever" became contagious, and people would take mini breaks with him, sitting on the floor and playing. As their business grew and they moved into a larger space, Brooke's boss ensured River would always be welcome there as well.

He wouldn't sign the lease except with a clause permitting River to be on site, telling Brooke, "We will never have a space where River can't come." He then doubled down on his declaration and River now enjoys his name frosted in glass right under Brooke's. And while Brooke's expertise lies in orchestrating cohesion between various departments, River should probably be named assistant to the head of human resources. Brooke says

River is very respectful to people, instinctively delivering what they need.

River's way of being in the world isn't just his day job. It's his way of life, which improves the experience of everyone he encounters. He breaks down barriers that might otherwise exist between disparate people who discover a shared love of animals. He helps people focus on the moment, slowing down to make connections. He helps bridge the divide between the office and home life. In this, he is a small dog with a mighty presence.

Charlotte and Amy

Like River, Havanese Charlotte spends some of her best days at work. But unlike River, she doesn't have far to go. Her mom, Long Island therapist Amy Baron, has her offices connected to her home and because Charlotte grew so accustomed to having mom around during the pandemic, Amy decided to allow Charlotte into sessions when in-person therapy resumed. Amy, of course, asked clients first whether Charlotte could be present, and their answers have been nearly unanimously "yes, please." So now, Amy and Charlotte often act as a dynamic team, Amy providing individual and couples therapy, and Charlotte supplying an almost intangible but vital add-on: no-strings-attached, non-judgmental love and affection.

As I spoke with Amy, I was reminded none of us would hate Mondays if we adopted Charlotte's attitude toward our day jobs. It takes only one invitation from Amy — "Come on, let's go to work!"—and Charlotte hops off her perch and races, without prompting, to the office door. Amy laughs, saying it seems to be something Charlotte puzzled out on her own — "work" happens in the "office." She wears her harness and leash, which Amy uses to signal to Charlotte there's work to be done. Once in session, Charlotte always takes her cues from patients.

Much like River, Charlotte "just *knew*" most of what she needed to become an indispensable part of office life. Amy shares that on day one, "Charlotte understood her place." She greets clients and can quickly determine whether they need her. Or not. If they just want a quick greeting, Charlotte respects that and then she'll settle in her bed on the floor.

More frequently, though, Charlotte knows a patient requires a bit more. In these instances, she will perch on the back of the sofa, nestling her head on a client's shoulder as they settle in for an emotional session. What awes Amy most is Charlotte's deeply empathetic and almost intuitive connection with humans.

Amy relates a particular couple's therapy session in which one spouse began sobbing. Charlotte, who was napping in her bed, woke from a sound sleep, jolted into action, and promptly leapt onto the crying woman's lap. In that moment, Charlotte offered empathy the woman's partner couldn't muster and the physical connection that Amy, as a therapist, couldn't appropriately provide. Charlotte now seems to watch for crying humans and has often repeated this type of soothing. Indeed, I wonder if perhaps Charlotte's small act of giving spontaneous, generous comfort to someone in pain might soften the tensions between couples. Sometimes, it takes a small animal to awaken our better human natures.

We know companion dogs provide emotional support humans often cannot, and we know companion dogs likewise bridge the divide *between* us humans, acting as a type of social WD-40. Small dogs do all of that and in a package that can more readily accompany us wherever we go, providing more immediate warm interaction. And since we've already proven small dogs are dogs, we know our petite canines can provide the same healing as their big dog counterparts. They do so in a wonderfully sized body that's easier to welcome into modern homes and workplaces. Small dogs de-stress, uplift, and heal us–they balance our perspectives and our mental health.

CHAPTER SIX

SMALL DOGS ARE EASIER ON OUR ALLERGIES, OUR HOMES, AND OUR PLANET

In our last chapter exploring how small dogs improve our health and happiness, we frequently mentioned small pups are perfect for such work because they're more readily admitted into health care facilities, businesses, and homes. The reason is a simple and practical one: Small dogs are less likely to cause allergy issues, they're easier to clean up after, and they cause less damage to our worlds.

DOGS AND HUMAN ALLERGIES

When it comes to the question of small dogs and allergies, it's important to know one thing: Puppy mill marketing to the contrary, no dog is "hypoallergenic." Although allergists once believed dogs who shed are more likely to trigger human allergies than low-shedding dogs, recent research reveals no evidence to support that hypothesis, except in the case of curly-coated dogs. People generally aren't allergic to a dog's natural clothing—whether such body protection comes in the form of hair that grows or fur that sheds. [101]

Instead, the guilty parties causing the sneezing, the wheezing, the swollen, bloodshot eyes, the sinus congestion, and the general misery are likely the proteins in our dogs' systems: their flaking skin or dander, saliva, urine, and poop. And obviously, all dogs shed these proteins. Thus, regardless of whether a particular dog sheds hair or fur, that dog will produce an allergic reaction in some humans.[102] Accordingly, the breeders bragging about their non-shedding poodle-crosses as "perfect for allergy sufferers" are engaging in misleading hyperbole. Perhaps they're operating on outdated science. Or, perhaps they're ignoring science altogether and employing their best marketing tactics. Either way, there's currently no such thing as a hypoallergenic dog.

SMALL ALLERGY BENEFITS TO SMALL DOGS WHO DON'T SHED

For most of my life, I remember being told that dogs who shed have fur, and dogs who don't shed have hair like ours that grows. Apparently, however, that's not a terribly helpful distinction. In fact, from a purely legal standpoint, it's not even accurate, as "dog fur" is defined by statute as including the pelt or skin of *any* Canis familiaris.[103]

For our purposes, we'll leave behind the hair/fur debate and focus on how both shedding and size impacts a dog's likelihood of sending some of us into sneezing fits and worse. Low-shedding dogs (like Havanese, poodles, Yorkies, shih tzus, bichons) shed minimally, whereas, obviously, dogs who shed are constantly leaving their fur everywhere. And on everything. Accordingly, the sheer *amount* of shedding fur spreads allergens because fur is covered in dander and saliva. And the fur left behind on our floors and furniture spreads throughout the house.

Thus, small dogs who shed can still create major problems for allergy sufferers. But with large dogs, the hair, skin, and saliva are prolific. A shedding dog's fur will also carry pollen in from the outdoors, and large dogs have more surface area and more shedding fur for such pollen to hitch a ride on.

For these reasons, small, non-shedding dogs are the best bet for people with dog allergies. Although they leave hair in their wake to about the same degree as we humans do, they aren't getting up from the sofa and leaving a chalk line shadow of fur behind as our goldens used to.

With our goldens, especially, hair everywhere became a fact of life.

Wispy undercoat floated in the air, landing on, and clinging to, the most inconvenient spots: my computer screen, our morning coffee, a guest's glass of really good red wine. Leaving for work in the mornings meant keeping an industrial lint brush in our bedroom closet to remove dog hair after dressing, and then another lint roller near the front door. Romeo specialized in a goodbye brush-up, wherein his overflowing feelings required he press his entire self up against our legs. So, we'd lint roll ourselves quickly after his morning parting ceremony, and then for good measure, we'd store another lint brush in our car to roll ourselves down one last time before walking into work.

Battling dog hair on one's person, especially when one's family includes big dogs, requires unrelenting vigilance. And forget about getting it out of your car mats, short of using jet-fueled, industrial vacuum cleaners. However you look at it, unless you're comfortable living in a fur-lined home, keeping up with the hair from a big dog is never ending and time consuming. And all that big dog hair carries dander and saliva.

Yes, the dander/saliva issue exists with small shedding dogs, but their lesser square footage means they're spreading much less sneeze-inducing protein. Not only are small dogs less likely to trigger allergies, but multiple studies also reveal newborns exposed to any dogs are much less likely to suffer allergies in the first place. So small dogs can both prevent and suppress the potential for allergic reactions.

Clearly, for people with allergies, small dogs have less adverse health impact. Small dogs who don't shed are even better, because they won't be leaving their pollen- and dust-laden fur all over our homes. That's especially true if we're diligent about using earth-friendly towels to wipe them down at the door and giving them weekly baths. And since we're on the topic of cleaning, let's explore how much your house and your schedule will thank you if you have small dogs.

SMALL DOGS EQUAL CLEANER HOMES

Our new hardwood floors were only weeks old when "Tom",[†] one of our kids' childhood friends, arrived on our back deck and let himself in through the sliding door. Out in the yard, he had picked up one of Romeo's tennis balls, thinking—with typical teenage boy reasoning—that

---

[†] Names have been changed so "Tom" doesn't feel bad, in case he reads this book.

indoor fetch with our eighty-five-pound golden boy was a good idea.

In true retriever form, Romeo smelled that rubbery tennis ball from three rooms away. He tore through the house, skidding to a halt in front of Tom, using his nails by way of brakes and my new hardwood floors as brake pads.

Hardwood floors weren't meant to act as brake pads.

The new finish hadn't fully cured, and the twelve-inch gouges left by Romeo's crash-landing resembled claw marks from some werewolf horror movie. I silently wished it were a horror movie. At least then I could stand up and walk out of the theater, leaving the gory damage behind.

After that day, the gouges gradually became more pronounced, as dust and dirt clung to the crevices where the new finish had scratched off. You couldn't mop it or scrub it out. The only solution would have been to refinish the floors.

Every time I walked into our freshly remodeled home we'd saved years to fix up, my eyes were drawn to the arced black claw marks in front of the deck door, where the afternoon light mocked me, spotlighting every gouge. I'd sigh as Romeo galloped to greet me, and I'd bury my face in his neck, reminding myself to stop caring about the stupid floors.

But I still cared.

Strike all that.

At least six times now, I've typed up and then deleted my memory of Romeo's destruction. Every time I begin writing it again, my stomach tightens as my words morph into complaints aimed at the beloved big dogs who've been my indispensable friends for the past fifty-plus years. These are the canines who developed my character as much as any human, teaching me responsibility, leadership, patience, and presence.

I always tried to be the same type of friend to them, embracing their doggy natures even when said nature dictated mayhem in my life. But the truth is, although I've adored dogs all my life, and most of those canines topped sixty pounds, it wasn't until I adopted small dogs that I learned the easy joy of canine life. One major reason for such simple enjoyment is the fact that small dogs are so much easier on our homes, our cars, and our clothes. They're quite simply cleaner, easier *to* clean, and cause less damage. But sometimes, none of that matters.

When Big Dogs Make Sense

When I was a child, our lives revolved around the outdoors; we were on the lake every weekend and on the farm at home during the week. Our cars, our clothing, our furniture, and our floors were all built around the idea that adults, children, and especially big dogs, would track much of the outside, inside. As a result, it was not only okay, it was encouraged (and sometimes required), that the dogs and the kids were outside—in rain, sleet, or snow—running, playing, working, and becoming wet, muddy, and messy in the process. In short, we had a life ideally suited to not merely accommodating but truly enjoying and helping big dogs thrive.

And it didn't hurt that my father was obsessive with the vacuum cleaner.

Many of us no longer have such lifestyles. And while countless big dogs live perfectly amazing lives with their responsible pet parents, it's equally true those parents often find themselves surveying their gouged floors, mangled woodwork, or torn furniture. They shake their heads while sighing, "This is why we can't have nice things."

But it doesn't have to be that way. We can enjoy all the dogginess we want *and* have nice things at the same time. Enter the small dog, stage left.

When Small Dogs Make More Sense

A pup under twenty pounds doesn't put measurable wear and tear on hardwoods or furniture. They don't weigh enough to scratch the hardwood floor finish or rip the upholstery. And if you're lucky enough to own a non-shedding small dog, you've hit the canine lottery, as you can invite them on the sofa with you and afterward not have to vacuum it to within an inch of its life. If your tiny pup *does* shed, you can keep a small blanket on the furniture and pop it in the wash, rather than worrying about a large sofa protector demanding commercial washing.

Neat? Or Clean?

Let's be very clear here though: "Clean" is a relative thing. To some people, "clean" means gleaming hardwoods and pristine, Zen-like surroundings free of clutter or indeed, any real evidence of human presence. If that's the clean to which you aspire, you're probably better off just playing catch with your neighbor's spaniel. Or winning a life-sized stuffed Lab at the annual summer's-end carnival and calling it a day.

In the cleanliness contest of small dogs vs. big dogs, small dogs *do* win, but they're far from perfect. As I write this, I count ten dog toys dragged unceremoniously from their designated toy tent and scattered throughout the first floor. A favorite stuffed animal has joined the pillows on the window seat, and a dirty sock, and what might possibly be underwear from the laundry room, is draped, embarrassingly, on the carpeted stair runner in the entry hall in full view of my front door.

And me? I'm incredibly relieved every major design magazine has declared the age of karate chop pillows "has to end" (though someone should alert Instagram), because apparently the dogs agree. Every time I straighten a pillow or strategically arrange a throw, cue the wrecking Havanese: One hops on top of the pillow to settle in for a nap while the other tugs on the blanket underneath her, attempting to dislodge both the throw and her sister from her perch.[u] Invariably, dogs, pillows, and blankets all end up in a heap on the floor.

Clearly, small dogs are not *neater* than large dogs. And, in fact, they are probably somewhat more prone to strewing clutter. But small dogs are infinitely easier if you enjoy a clean house. They're a snap to bathe, they shed a mere fraction of the hair their larger counterparts produce, they track in much less debris (and are much easier to wipe down), they rarely drool, they don't make dirt-collecting gouges in the hardwoods, and they don't step in their water bowls.

And yes, my guilt is tapping me on the shoulder as I list what always drove me crazy about my beautiful golden

---

[u] And here's another example of small dog smarts: Any dog who understands that pulling on a blanket underneath her sibling will knock said sibling off a favorite spot possesses a plotting mind deserving of respect. And a bit of careful wariness.

retrievers. Yet, that guilt is met with admitted relief for the greater work-to-joy ratio our small dog pack provides.

While small dogs stand guard at windows and sliding glass doors just like their big dog counterparts, they don't drool on them with the same enthusiasm. With our sweet golden boy Romeo, washing the glass door to the deck was a daily chore, requiring heavy duty window cleaner and a muscled arm. If we skipped a day, the glass sported slobber streaks, making it look like we hadn't washed it in months.

Phoebe and Scout are also vigilant window watchers, yet it takes only a dry microfiber cloth to remove their tiny nose prints. Likewise, if they run headfirst into the screen door they assume is open, Phoebe and Scout just bounce off the screen. Romeo? He took out the entire door once when he barreled into it at full speed. Small dogs do the same things our big dogs do, but their footprint is literally and figuratively so small as to cause barely a dent.

Similar benefits exist in the kitchen, which was Romeo's favorite place to spend time with his humans. In the kitchen, small dogs don't leave puddles of drool or bruises on your foot when they accidentally step on you while seeking to inspect the night's menu. Small dogs have small bowls they tidily eat from, not large troughs they bang loudly and messily against the baseboard moldings. With Romeo, I spent at least thirty minutes weekly scrubbing my walls and woodwork clean of the food bits, saliva, and hair he'd let fly during his enthusiastic love orgies with his food dish.

And yes, of course, I'd gladly devote that time again if I could have him back. But that doesn't change my gratitude for the recaptured hours and new sense of calm in my home now we've downward-sized our dogs.

DOWNWARD SIZING DOG

SMALL DOGS ARE EASIER ON THE YARD

Just as small dogs are easier on things *inside* your home, they're easier on its exterior as well.

If you have a yard, and you've ever had a big dog, you know the two don't get along. Or perhaps it's more accurate to say the connection between your lawn and your big dog is not a mutually beneficial relationship. It's more of a one-sided, unrequited-love-type coupling.

Because, in truth, big dogs *love* the yard.

They love tearing racetracks in its grass, digging holes in its expanse, and burning pee stains in its green carpet. When they're done, if they're big dogs, the yard looks nothing like the rolling green acreage you nurtured it to be in your mind. No. Instead, it sports patchy bald spots, dangerous depressions, and large, golden-brown tufts of dead grass.

Of course, if you adore big dogs, there are solutions to prevent these crimes against your piece of paradise. But they'll cost you, either in time or in money. Or both. You can do what my dad and his beautiful wife do: They follow their dogs around with a garden hose nine months out of the year and they water the grass where (and when) their dog pees.[v]

If you're not so extreme, you can install fencing to keep your big dog's damage confined to a reasonable area. You can even bar your dog from the yard altogether and opt instead to take her for walks everywhere. You could move to the desert where grass isn't a thing or convert your yard

---

[v] More full disclosure: With two small dogs, and one who likes to pee OVER her sister's pee, I do get spots. Now that Scout has taken on the trait of covering her sister's scent with her own, we've resorted to hose-following too. But still, any spots we miss are miniscule compared to what we had with our big dogs.

to an eco-friendlier landscape (we should probably all be doing that). And, since not all large canines dig or tear up the grass, maintaining a gorgeous yard and having big dogs *can* be done.

But consider this: Small dogs and gorgeous yards (just like small dogs and clean-ish homes) go hand-in-hand, because, generally speaking, little breeds don't possess the physical heft to cause much damage. Thus, for example, even if you have a pup whose genetics predispose her to digging, her excavations will cause less damage than a big dog digger. If your little dog loves to dash around in circles, his light weight prevents him from tearing a permanent trail in your yard. And the unsightly grass-burning? Instead of basketball-sized spots, the patches will be so small, you'll likely not notice them.

Finally, with small dogs, the outdoor waste disposal is likewise less time-consuming and infinitely less gross. I won't stoop to all the gory details, but boy, picking up after a large dog is truly just odious, especially if you're feeding kibble. It just is. Whether you're using a tool or a bag, it's way too much to handle, and it's so unwieldy it's hard to get it all. But with a dog under twenty pounds? It's only marginally disgusting. Small dogs are all the fun and half the work.

SMALL DOGS ARE GREEN(ER)

Small dogs are easier on our personal environments. But what about our overall global ecosystem?

When it comes to dogs and "the environment," I struggled with whether to discuss the relative eco-friendliness of small dogs. The reality is we all pollute the planet. Unless you're living on the land and off the grid, nearly every human action has a negative environmental impact. And in spreading blame for environmental damage, we humans tend to judge one another unfairly based on our lifestyle choices.

Electric car drivers fault SUV owners, who point fingers at lithium battery users, who blame people who use plastic bottles, who criticize those who still use paper towels and napkins, who censure the vegans with more than two children, and so on. Given dogs in general have recently been targeted for their adverse environmental impact, it's a bit misleading to say small dogs are more environmentally friendly.

And yet when compared to big dogs, small dogs *are* easier on the planet. Sometimes significantly so. In fact, a small dog's environmental paw print is so much daintier than his larger relatives in the dog kingdom, experts on sustainability often suggest small dog ownership to soften our impact on the planet. Little breeds consume less meat, produce less waste, and interfere less frequently with wildlife habitats.

THE FOOD PROBLEM

So first, let's talk about the food our dogs eat. The ugly truth is dogs (and cats) contribute to about thirty percent of the pollution attributed to America's meat-consumption.[104] And their waste, if not properly disposed of, likewise poses health and environmental dangers to our coasts and waterways. That pollution results from the "use of land, water, fossil fuel, phosphate, and biocides."[105] And that thirty percent figure doesn't even account for the massive environmental costs of transporting all that food.

To be sure, smaller dogs have higher metabolisms and their smaller food consumption isn't directly proportionate to their smaller size. But small dogs consume a fraction of the food big dogs eat. For example, if I were feeding Phoebe and Scout only their freeze-dried raw food, they'd each eat about a cup a day. My golden retriever Romeo, who weighed at least sixty pounds more than they do combined, would need about eight cups a day. That's one large dog needing *four times* the food two small dogs need.

WHAT GOES IN, MUST COME OUT: THE WASTE PROBLEM

Dog waste, if not properly disposed of, poses significant health and environmental dangers to our coasts and waterways. And not just because of the E. coli, giardia, salmonella, worms, and other dangerous toxins any mammal waste can contain. Because too many of us, for the most part, feed our dogs processed, high-nutrient kibbles, their waste is rich with nitrogen and phosphorus.[106] And those nutrients can lead to algae blooms that damage or outright destroy our waterways. (It's one reason we feed our pups their highly digestible freeze-dried food—it drastically reduces the amount of waste the girls produce. That, from an eco-friendly standpoint, is key.)

It doesn't take much waste to create nearly irreparable environmental damage. According to my state of Rhode Island, the federal Environmental Protection Agency reports the waste from just one hundred dogs in the space of a weekend can temporarily close a bay to swimming.[107]

In my little village surrounded on three sides by water, there's a population of roughly twenty thousand people. On my street alone, nearly every other home has a dog, and most of us have two. You can't go outside without seeing someone walking their dog. It doesn't matter when—in the early morning before dawn when only the gulls are crying, or late at night when I watch for coyotes and fox, or in midday when those working from home or retired take their lunchtime strolls.

We may not have one hundred dogs in my immediate neighborhood, but we have enough to create a major problem. Fortunately, all but one or two of us are vigilant about picking up after our dogs (and, boy, would I love to have a chat with the human who isn't). Whether we all know what to do with the waste once it's picked up is another story. It's something we might want to chat about at the next neighborhood get-together. In the meantime, I am happy to know my small dogs are (1) eating a relatively small amount of food and their bodies absorb most of the nutrients, (2) they're producing much less polluting waste, and (3) we abide by a "pick it up immediately" policy preventing waste from draining into our waterways.

Strictly speaking, all dogs damage the environment. But relatively speaking, small dogs are so much better for the planet.

CHAPTER SEVEN

SMALL DOGS TRAVEL WELL

So I'm aware of the irony here: I'm moving from environmentalism to talking about travel, which is apparently one of the biggest offenders, comparatively, in creating greenhouse gasses.[108] Since this is a book about embracing all small dogs as family members, we won't dwell on that point, although we feel compelled to disclose that apparently eight percent of our carbon emissions stem from tourism.[109] However, traveling with your small dogs certainly doesn't *add* to climate change, given the pups don't take up any seats.

Now, back to our regularly scheduled program.

As long-time golden retriever parents, we were, like all big dog owners, practically barred from traveling with our dogs. We live on the East Coast. Most of our family does not. We have a child in southern California, a child who splits his time between New York City and LA, and my extended family calls Michigan and Arizona home. Most of our vacations comprise jetting from East to West Coast or to the Midwest and back again. That meant leaving our goldens behind.

You really can't fly your big dogs for casual week-long vacations unless you're into danger and anxiety. And even short weekend getaways become problematic when hotels impose size limits on pets. So big dogs usually get stuck at home with a sitter or sent to a kennel.

Small dogs, once again, are good to go. While you'll incur airline charges for the privilege of letting your pup sleep under your seats, and you'll lose out on the convenience of carry-on luggage (because that's what they count your pup), those sacrifices are trivial. There's nothing like bringing your entire family with you. In this chapter, we'll look at the benefits of traveling with your small dogs, not the least of which is the joy they spread to others as they reach their destinations, whether you arrive by car or plane.

Have Small Dog, Will Travel (by Car)

Growing up in Michigan, our jet-setting comprised piling inside a wood-paneled station wagon every weekend for the five-hour drive to our cottage on Lake Millecoquin in the Upper Peninsula. My parents would arrive at school on Fridays just as the dismissal bell rang, and we'd be the envy of all our elementary school friends when my dad opened the tailgate to load us in.

It was in the pre-seatbelt-required era (that was a thing until 1985 in my home state), and both my parents had a knack for knowing what kids found fun. My dad was particularly skilled in the art of comfortable car travel, which usually involved a clubhouse-camp vibe.

Small Dogs Travel Well

In the back of the wagon, he would arrange a foam mattress with sleeping bags on top, and my mom would pack snacks, books, and games for us. Our Irish setter Meg would be waiting for me, already having staked out sixty percent of the available space. There, my sister and I would happily[w] spend the entire trip playing cards or roadside bingo, camped out with Meg's head in our laps or, when it became dark, lying down and trying to sleep, using Meg as a living pillow.

It's a miracle we didn't all die in a tragic car accident (and given the two-lane highways in Michigan in the seventies, that was a significant possibility). But we didn't, and at the time I believed traveling without seatbelts, with a big dog lying where she pleased, was not only no big deal, it was heaven.

Of course, my view likely bears little relation to the reality my mom and dad faced. I was just the kid in these scenarios, and I didn't have to deal with the mud or dog hair left behind when we disembarked from our luxury travel pod. My parents did.

And it wasn't just mud and hair. My beloved Meg had a destructive side. She was infamous for having destroyed my mom's favorite childhood poetry book. Fifty years later, she's become legendary, as my mom, in her eighties now, reminds us at every gathering about "that dog" and her destruction. In a separate crime, Meg ate the entire arm off my mom's newly upholstered chair. And I'd bet money that

---

[w] I lapse into unabashed fiction here—we were famous for bickering, and quite skilled in the art.

if she never *purposefully* gnawed on the car seat leather, certainly her nails poked a few holes in it.

It was true then, and it is true now: Unless you truly don't care about the costly vehicle you've invested in to ferry you from Point A to Point B, know there's significant work and expense involved in protecting yourself, your passengers, and your car from a large canine.

First and foremost, there's the safety aspect of car travel with dogs. With small dogs, one foam car seat can be buckled in quickly, with the dogs tethered inside the car seat. Even when our Havanese's car seat isn't already in the car, it takes mere minutes to secure, including fastening Phoebe and Scout's harnesses to the inside. Three minutes, tops. Done. We're off in no time. In contrast, riding in cars with big dogs requires much more forethought, equipment, and, often, physical strength.

You have choices, to be sure.

Is your big dog riding in the back seat? Will he jump in himself? If so, are you prepared for your seats to be gouged or worse, for him to hurt himself by landing awkwardly? Or will you sacrifice your clean clothing, and possibly your spine, to lift him up?

If he's not riding in the back seat, will you keep him in the cargo area of your SUV and sacrifice your rear view with the large crate or the back-seat barrier? You don't really need to see what's behind you, right? And again, is your big dog young and healthy enough to jump in himself? Or will you need to lift him or grab that heavy, awkward dog ramp so he can walk up? Whatever option you choose, it involves potential damage to your car, your clothes, and, most importantly, your beloved dog.

But wait, there's more.

You can ask yourself the same questions about unloading your dog when you arrive at your destination, again when you leave, and then again when you arrive home. And, when you *do* arrive home with your big dog, you have some additional choices: Do you take time to lug out and put away the heavy seat covers or the large crate in the back? You know if you leave the crate or bedding, your car *will* smell like a dog. I love dog smell as much as the next normal human, but I like the smell on my dog, not in my car. And did I mention the smudged windows needing de-slobbering?

Even if we're very generous and assume it takes only five minutes to secure your big dog in your car and only ten minutes to load and unload the equipment you need to do so, you're adding thirty minutes to your trip. Do your days include random half hours you can afford to lose?

And that's often why we sacrifice time with our dog and do what too many big dog owners find themselves having to do: They leave their dogs home. It often happened with Maddie, Haly, and Romeo, our golden retrievers who just weren't companions you could bring into a store for a quick minute. True, these beloved family members spent hours upon hours at the ball fields with us watching our kids play soccer, softball, and baseball. But our busy lives required us to leave them home for pretty much anything else. And we were doing "anything else" almost all the time.

We had to leave them behind.

A lot.

When twelve-pound Phoebe and ten-pound Scout became part of our family, life suddenly opened up for all of us. Rather than leaving the dogs at home, "home" became wherever we happened to be...because we just brought the dogs with us. Since they're not only small but also considered "hypoallergenic,"[x] and more importantly, very well-socialized, we became quite brazen in asking, "You don't mind if we bring the dogs, do you?"

To be sure, we're certain some people *do* mind, very much, but they never say "no." And then once they meet Scout, the charmer in the family, they don't want us to leave. Actually, they're fine with *us* leaving; Scout? Well, her they want to keep. It's a truism: Even long-time small dog haters fall in love with Scout and volunteer, repeatedly, to take her off our hands. When we reject their offers, they invariably resort to picking her up and feigning absent-mindedness as they pretend to leave with her.

Not only are small dogs easy to cajole a "plus one" social invitation for, they're almost always permitted—and often even welcomed—in many stores. We can plan an entire day with the dogs knowing they'll be fawned over wherever we go, whether that's an outdoor lunch, shopping, or running into the local bank. Although most of these places allow large dogs as well, the reality is bringing a Dobie or border collie into Pottery Barn is going to result in some breakage.

A typical large dog's tail is at a perfect height for sweeping those wine decanters off the shelves. And with a large canine, you can't just swoop them up into your arms and out of harm's way. With a small dog under twenty pounds, you can avert potential disaster—whether that's an overeager toddler barreling toward you or a precarious plate resting too close to the edge—simply by scooping up your dog. It's a win-win.

The evidence is clear: If you're a person or family in constant motion, the so-called traditional spaniel or

---

[x] Remember, in the previous chapter we warn there's really no such thing as a hypoallergenic dog.

retriever may be the opposite of what's best for your family and the dog. Small dogs are easy to transport in the car and easy to accommodate when you arrive at your destination. They're perfect for short or long road trips.

Have Small Dog, Will Travel (by Plane)

But enough about cars. If your particular brand of constant motion involves frequent air travel, and your dog is too large to fit in an approved carrier under the seat in front of you, then your big dog will have to be left behind. That is, unless you're willing to check her as baggage or transport her as cargo. Both options are expensive (and during COVID pandemic conditions, most airlines suspended checking animals as baggage) and unless your dog is unusually chill, being baggage or cargo is at best terrifying for your pet and at worst life-threatening.

In addition, airlines frequently cancel cargo and baggage service for pets because of temperature extremes. In fact, many popular airlines refuse to accept dogs for shipping cargo, indicating pets may travel only in the cabin, and only if they fit comfortably under the seat in a carrier. Of course, thousands of large dogs *do* fly in the baggage area or cargo hold every year. Most of the time, it all works out okay.

Sometimes, it doesn't.

According to the U.S. Department of Transportation Air Travel Consumer Report, airlines in the United States transported over four hundred thousand animals in 2019. They reported ten animal deaths and seven animal injuries.[110] Statistically speaking, that's barely measurable. But our dogs are not numbers to us. If our dog were one of the .000025 who died on a flight, the loss would be immeasurable.

Besides the safety concerns involved in sending your large dog as cargo or baggage, you're looking at significant cost. A recent check shows two major airlines charging nearly $450 one-way to transport a fifty-pound dog in their cargo hold. Add the cost of the veterinary visit and health certificates required to ship a dog, and most people opt to leave their large pups at home with the pet sitter or in a boarding facility.

We used to do so all the time. When we left them behind, our golden family members were well-cared for, and, while not exactly happy, they were safe and comfortable.

And it was fine.

But "fine" isn't what we're going for here. Most of us just don't want to leave our furry family members home. We want them to enjoy our vacations and our travels. We want to enhance our own experiences by seeing new places through their eyes as well. We want to enjoy the pleasure of another family rushing over to ask, "May we please pet your dogs? We had to leave ours at home." Bringing your small dog on vacation is, quite simply, a transformative experience.

Flying with a small dog is a shockingly simple thing to do. My sister and her shih tzu-poodle mix Abigail spent fifteen years of Abbie's life traveling by plane from Arizona to Michigan or Rhode Island. They'd make the trips at least several times annually.

My first experience with bringing a pup in the plane cabin occurred when I flew to Ohio to pick up Scout from Karen Warncke, our wonderful Havanese breeder. While it was only a two-hour flight, I stressed for weeks in advance about whether ten-week-old Scout would have anxiety on the flight, or whether a fellow passenger would be upset about sharing airspace with a dog. To my great relief, the opposite of my fears occurred: Scout slept soundly through nearly the entire flight, and the only complaint from others around me was...that she slept soundly through nearly the entire flight. They wanted me to take her out and wake her up so they could absorb her adorableness.

Our next flight with the pups occurred a little over six months later, when we flew with the dogs to Arizona to visit my mom and sister. Mom had been facing health issues, and she wanted to see "her puppy" Scout, as she had been there to help me pick her up in Ohio. We'd never taken a long vacation with the dogs, but we were willing to try because we knew how much my mom would appreciate it. We booked our own flights and then the extra tickets for Phoebe and Scout. I bought the approved under-seat carriers and we braced for what we assumed would be an un-relaxing cross-country flight. What we found? Other than the stress over the dogs refusing to use the airport pet relief stations (does any dog actually agree to use those?), air travel with small dogs is really fun.

The dogs were impressive pup ambassadors in the airport, strutting with confidence in the pre-COVID crowds and throwing off love to anyone who stopped and asked for it. Scout's curiosity around several wheelchair-bound adults created a small crowd watching as she danced on her hind legs, attempting to see what treats they might have in their laps. And the children...so many kids in the airport gravitated to the girls we began worrying we'd suddenly find ourselves with a spare child. Sometimes, the kids didn't wait to ask their parents' permission and just toddled off to say hi to us.

DOWNWARD SIZING DOG

Once on the plane, Phoebe's carrier slid under Ron's seat and Scout's under mine. Here, we did run into some trouble, as Scout became indignant about being zipped away. For a good half hour, she pummeled the front of the kennel with her paws and whined. We used the trick my sister taught us from her many years flying with Abby: We unzipped the carrier's side and put our foot inside the kennel. Within ten minutes, Scout was snoozing against my sock, and the six-hour flight passed without further issues.

Although flying with small dogs is relatively easy, the bathroom thing can be a challenge, especially assuming your dogs aren't trained to go on pee pads. Many city dogs and dogs unable to brave bad weather have an advantage here, and I admit my insistence on treating my small dogs like every other big dog is pretty stupid here. If I had trained them on pads, then they would go at the various pet relief stations now at every airport or I could use a pad in any restroom.

My dogs looked at me like I was nuts when I suggested, begged, bribed, and cajoled them to "go potty" while pointing either to fake grass or a blue pad on the floor. They would both rather die, I think, than go to the bathroom indoors. It's just not a thing in their world. So, one source of major stress on this trip was worrying about their comfort as they "held it" for a good six hours. But they did, and neither seemed to suffer any ill effects because of it. And really, they both hold it every single night for well over eight hours, though in those situations their entire system is in sleep-mode. If I were doing it over—their puppyhood, that is—I might train my small dogs on pee pads first, and then to go outdoors. Doing so would make cross-country flights a more stress-free experience.

So, let's return to our comparison. If you fly with your big dog, you can't bring him in the cabin (unless he's a service or emotional support dog), and then you're limited to the few airlines still permitting dogs to fly cargo. You'll pay dearly—in the range of $900 round trip—for the privilege of putting your big dog through a significantly stressful, lonely, and arguably hazardous ordeal. If you fly with your small dog, you'll always have her with you, she'll spread joy at the airport[y] and in flight, and it will cost roughly $250 round trip.

So, in a nutshell: have a big dog? You're not likely to fly anywhere with him. Have small dog? Will travel.

Getting from the Airport to the Hotel and Everywhere Else

If you're just taking a cab, a Lyft, or an Uber, you can easily find a pet-friendly ride—it's merely a matter of making a special request on the apps. And, of course, pet-friendly hotels have pet-friendly shuttles. However, if you're renting a car, you'll want to be a bit more vigilant to avoid being gouged with unnecessary fees. Some smaller car rental companies still charge an upfront fee for the privilege of bringing your pet. But should they? Aside from a little shedding, small dogs are much less messy than a two-year-old with goldfish and a juice box. With Phoebe

---

[y] Even more full disclosure (I tend not to hold back): Of the thousands of people we encountered in Boston and Phoenix airports, only one, a man who was angry because I couldn't move my luggage and my dogs out of his way quickly enough for his pleasure, had anything to say. His brutal "f$%king animals" comment deserved rebuke, but I looked at Phoebe and Scout, took a breath, and left him to his misery, sending up a silent prayer for any animals unlucky enough to be in his life.

and Scout, two non-shedding Havanese, they're paws-down less messy than children.

Most major car companies don't charge an extra fee, although some franchise locations ban dogs, so you need to check. In addition, at least one company requires you to use a car with a pet barrier. And all of them will charge you a cleaning fee if you return the car covered with hair or other evidence of pet wear and tear.

But if your dogs are small, you'll leave no gouged seats or slobbered windows. Whatever slight evidence the pups might create can be erased with a quick vacuum or wipe-down with a damp cloth before dropping the car off at the end of your stay. In contrast, if you're transporting a big dog in a rental car, you're much more likely to end up paying a hefty surcharge to erase the inevitable evidence of your well-loved but large, shedding, slobbering passenger.

THE TRAVELING WELCOME MAT: SMALL DOG-FRIENDLY STAYS

As you've probably noticed, in the twenty-first century, Americans promoted dogs to near-human status as family members. As a result, we expect them to be welcomed when we bring them with us on the road. And, since the hospitality industry is all about, well, being hospitable, it's responded remarkably well, making it very easy to find places to stay with our dogs. Whether you want to rough it in a campground, stay at a roadside motel, rent someone's Airbnb, or luxuriate in a five-star hotel, you can easily find a place—especially if your dog is under twenty-five pounds. In the past, you could Google "pet-friendly hotels" and quickly find a list of decent accommodations, if you were willing to pay the $25-$200 fee the hotel charged for your pet to stay. Now? Those fee-gouging days seem to be a thing of the past. Or at least they were before this inflationary period hit.

Today, it's probably more efficient to Google "hotels that *don't* allow dogs," to know what chains to avoid. Because most major hotels? They're not only dog-friendly, but they're also waiving the pet fee, styling doggy welcome baskets, and greeting you and your pup on the red carpet. They're even pouring bottled "still, not sparkling" water in the monogrammed dog dish. It's true: You can bring your under-twenty-five-pound dog along with you nearly anywhere you want to travel in America and not pay any additional fee to stay overnight. The same is not true for larger dogs.

Well-mannered, affectionate small pups spread joy on the hotel or resort property wherever they go. (In fact, our Phoebe and Scout are such popular attractions sometimes I think the hotel should pay *us*). Unlike large dogs—even perfectly obedient ones who can accidentally track major mud, or knock over a plant, or break a vase with a sweeping tail, or back up on a toddler—properly trained small dogs simply have small footprints.

And we mean that both literally and figuratively. Encounter a muddy path? Scoop them up and carry them over the mess. Suddenly find yourselves in a crowd? Pop the pups in a carry bag and the problem's averted[z]. Picking up their poop? It's the size of a baby carrot and almost as easily disposed of. A small dog's way of being in the world is a perfect mix of canine power and convenient packaging.

---

[z] Scooping up your small dog is not something you should do frequently, as discussed in the chapter entitled "But What About Small Dog Syndrome?"

When Do We Leave?

Thus, when it comes to travel, small dogs allow you to keep the *entire* family together almost effortlessly. Phoebe and Scout are living proof, even though we were hardly seasoned dog travelers. Yes, we were incredibly nervous for our first major trip out West with the girls, but they behaved as though they traveled like this every weekend. They were wholly unfazed in the airport and on the plane. When we landed, they trotted through the airport like authoritative police dogs. In a matter of hours at the resort, they owned the place.

Dozens and dozens and dozens of people stopped to say hello—not to us, but to the pups. We think Phoebe and Scout single-handedly lowered the anxiety of hundreds of people throughout Boston and Phoenix. And they did the same for us. Because we could bring our small dogs along, our vacation days were more fun and our nights more relaxing.

What was that saying? Have small dog—will travel.

Small Dogs Have Built-In Babysitters if They Must Stay Home

Despite their near-perfect travel skills, sometimes, of course, you must leave your small pup behind. In these cases, the small dog/big dog debate isn't a fair fight: Well-trained small dogs beat well-trained big dogs every time. With Phoebe and Scout, for example, we have family arguing over who babysits. Good friends frequently drop hints we should probably go away for the weekend so they can watch the girls. Their petite stature and damage-free portability make them shockingly popular guests.

Small Dogs Travel Well

In our golden retriever days, we never had such offers. Friends and family may *claim* they don't mind dog hair and large-dog demands that include fist-sized poop patrol, but they're certainly not lining up to invite your seventy-pound dog to stay for the weekend.

When we had our goldens, visitors would say, "Goldens are just the *best* dogs," but unlike now, when everyone tries to stuff Scout in their bags to take home, they'd never try to borrow our dogs; instead, they'd steal our lint rollers on their way out our door. When we'd start making calls to find dog sitters, our normally available friends and family would develop raging cases of jam-packed calendars. To be sure, we had one friend of our daughter Caroline who loved Romeo like her own, but her calendar was legitimately packed all the time. Consequently, we'd be forced to use local kennels.

In contrast, people welcome small dogs into their homes for short stays, especially when their temporary guests are non-shedding. Family dog sitters aren't worried about the scratched floors or punctured furniture caused by larger canines, and an anxious small dog missing her family is much easier to calm than a seventy-pound bundle of nervous canine energy. In fact, small dogs are so welcome they can put dog grandparents in awkward situations.

"Diane,"[aa] whom we'll get a better introduction to in our chapter on small dogs/small kids, illustrates the point perfectly. Diane's mom adores Diane's two Havanese, Charlene and Jack, and has extended a standing offer to babysit the dogs. That same offer is glaringly absent for

---

[aa] Names and breeds have been changed to preserve family harmony.

Diane's sister, who has a rambunctious and powerful German shepherd.

And to the sister, that feels like favoritism.

But the reality is it's wildly unrealistic to expect someone's unbridled excitement at the prospect of babysitting your big dog. Even when it's their granddog. Nannying a large, exuberant animal who's not your own is in the category of "helping someone move" or "coming over to paint a room or two." We do it, but we don't *volunteer* to do it, we don't *want* to do it, and we'll grasp at any reason to *avoid* doing it: "I'm so sorry, I would have loved to, but I can't. I've promised so-and-so I'd do such-and-such."

But with our little pups? Whether we're taking them with us or leaving them home, their benefits far outweigh their actual poundage. Small dogs are welcome pretty much anywhere one wants to explore. They spread affection and calming vibes wherever they go and to everyone they meet. They give us the best way to travel, where we can both escape everyday life and yet bring part of home with us. And when we must leave them behind, finding friends and family to care for them is usually a matter of choosing the best offer.

In downsizing our dogs, we may upsize our experiences not only for our pups, but for ourselves and those around us.

CHAPTER EIGHT

SMALL DOGS ARE LESS EXPENSE
(Unless You Spoil Them, but That's on You)

While traveling with small dogs (or providing care for them at home) is not cheap, on a day-to-day basis, small dogs weigh much less heavily on your wallet than their big breed brothers. Small dogs eat a fraction of the food large dogs eat, and feeding tends to be the biggest expense in caring for your pup. According to some estimates, small dog owners spend roughly half of what large dog owners spend annually.[111] And, while small dogs may appear more costly over their longer lifetimes, when you run the numbers, it's clear small dogs are much more affordable than large pups. As we'll see in this chapter, when we compare the proverbial "apples to apples," small dogs are a better choice when finances are a factor.

May, quite some years ago.

I take Romeo to the vet for his annual check-up and his three-year rabies shot. Ticks are teeming our property, and I shove aside my anti-pesticide stance to put him on tick preventative.

Bill? $175.

Early July, ten pm, that same year.

I am at the tail-end of my nightly ritual involving dog dishes, dishwashers, and coffee pots. My final job, before sinking into my soft bed and reading for fifteen minutes, was taking Romeo and Phebes out for their nightly trot. We lived on two acres among homes who likewise enjoyed such space, and then some. Evenings were quiet except for the chatter of crickets and frogs, and our long driveway lined with trees engulfed us in suburban wilderness, lit only by the waning moon.

On this night, there is no rain, but with the oppressive humidity, we feel drenched. Because Phoebe fears nothing, I have her on a leash when we venture into the yard at night. We'd been hearing Fisher cats and had been swooped down upon by large owls, so the leash keeps her close and safe. But at eighty pounds[bb] and with virtually no prey drive, Romey always trotted freely alongside us. Usually, he would inspect Caroline's car for interesting odors, and then he'd catch up with us at the driveway bend to over-mark Phoebe's scent with his own, in an instinctive, but uncharacteristic, show of alpha manliness.

[bb] Alert readers will note that Romeo's weight often changes, and it did. He tended to be thinner in the summer and put on the same winter weight his human family often gained in cold months.

Small Dogs Are Less Expense

But on this night, before Phoebe and I reach the driveway bend, Romey's violent sneezing stops us both, and we look his way.

I hear an odd clacking noise, and I look up to see Romey's form take shape between the pitch dark and the angled light of our garage lanterns. The harsh sound emanates from his jaws, which he's opening and closing as though he's eaten something disgusting.

I steel myself for the inevitable.

He hasn't rolled in or eaten deer droppings in two years.

Why tonight? I'm exhausted.

I call to him. But instead of running to us as usual, he rushes away toward the open mudroom door, looking afraid. Again, I hear his teeth gnashing (who knew, that's literally a thing), and I abandon the idea he's "just" rolled in something. If that were the case, he'd be proudly showing me his handiwork. No, something is wrong. My first thought is that he's been stung. Or maybe a porcupine has gotten him—although I had no idea whether such impressively armed mammals existed in Rhode Island.[cc]

Calling again to Romeo, I'm flooded with relief as he approaches, and my phone's flashlight casts an eerie halo around his uninjured face. But my relief quickly dissipates.

The heavy air in front of me does not.

It reaches me before Romeo does—a living thing, acrid and angry.

My brain screams, "No, No, No. No, No, No."

And then it sobs: "Skunk."

[cc] *And if, like me, you suffer chronic curiosity and were pausing to wonder whether they do exist in the East, here's the answer: Yes. But they're rare.*

DOWNWARD SIZING DOG

Anyone who's ever dealt with skunk spray knows that you don't have to be the one sprayed by it to end up smelling it, for weeks, afterward. When it's fresh, skunk spray clings to hands, hair, shirts, pants, shoes, the inside of your nose—everything it touches. And then, it doesn't wash out. The odor's top notes will, of course, succumb to water, but the scent's under-notes linger for what seems like forever.

I grab Phoebe, who loves everything about her big brother and seems ready to bury her face in his skunk-soaked face, just to check out his new scent. And as I back away from the pungent wall of polluted air, poor Romeo, trustfully expecting me to fix his predicament, wags his tail, rears on his hind legs, and tries to climb into my arms.

I step back, but Romey matches my every backward stride with a forward step, becoming upset and confused by my evasions. He circles us, crying, and we all begin an awkward dance to the garage. Once there, I drop Phoebe inside the mudroom hall, yelling to my family for help.

Finally facing Romeo, I see with dread he's suffered a direct hit, his face soaked and dripping with skunk spray.

That was at ten p.m. By midnight, Ron, my daughter Caroline, and I had done the best we could with him. We'd all spent the last two hours in the driveway soaking a bewildered Romes in a blood-red concoction of tomatoes, baking soda, apple cider vinegar, and Dawn. Thank goodness for Dawn. We should have used peroxide too, but we didn't have any on hand.

We stumble to bed. Despite five shampoos, Romeo still reeks of skunk. He's barred from our bedroom for the first time in his seven-year life.

The next day (Sunday): Emergency grooming appointment.

I find half-a-dozen open groomers, and one, the local big box pet store, can fit him in that day. Under less desperate circumstances, I use small, independently owned groomers, but here I must give praise where praise is due. The store groomer greets Romeo as though he smells of violets and sunshine and not acrid skunk. She kneels to calm his anxious crying and panting, and when I pick him up four hours later, his odor is detectable only if you press your nose to his face.

The bill? $125 plus an extremely generous tip. And another $30 to replace the tomatoes and restock the cleaning supplies from our cleanup the night before.

Two days later.

I notice a weeping injury between Romeo's eye and ear. Terrified the skunk had bitten him and somehow we all missed it, I rush him to our vet.

The diagnosis? No puncture wound, but instead he has infected hot spots and an ear infection for good measure. He's never had either issue, and it's likely from the skunk spray and our clean up. Our vet shaves him, cleans him up, and sends us packing with three medications.

Bill? $160.

End of July: Saturday night, around eight p.m. Still that same summer.

Romey begins vomiting violently, out of nowhere. He's lost what appears to be both his breakfast and his dinner, and it smells so foul. He'd suffered from bloat three years earlier, a condition in which the stomach twists on itself and causes death within hours. It tends to occur in larger, barrel-chested breeds, and it's common in male, middle-aged golden retrievers. That's my Romeo, exactly.

Although actual vomiting isn't a common symptom for bloat, he's clearly in distress—panting, pacing, and crying—even after vomiting five times. Ron and I look at each other. A trip to the state-of-the-art veterinary hospital in Rhode Island is never good, either financially or emotionally. They're incredibly talented, but we've never left there with actual good news. We reach silent agreement: We need to take him. This time, for once, we do get good news. Just a bad gastroenteritis. Two X-rays, two consults, and over two hours later, we bring Romeo home with anti-nausea medication and pumped full of fluids.

The bill? $776.

Total unexpected Romey expenses for May and June? $1,236.

From a cost standpoint, owning any dog is much like owning a car: After the initial expense, you feed them and take them in for regular checkups, but there are those inevitable, always inconvenient emergencies. The unexpected brake job that sets you back $600. The engine that dies just on the wrong side of the warranty period. The socks your Lab eats requiring $2,000 surgery.

Of course, small dogs likewise suffer illness and injury. And I possess the bills to prove it. But it's nevertheless true large breed dogs have much larger veterinary bills on average than do small breed dogs. I even have a direct comparison: Romeo's emergency vomiting/diarrhea bill at the emergency vet above was $776—and that was pre-inflation prices from over five years ago. I brought Phoebe in last month for nearly the exact symptoms at the same emergency hospital, for the same diagnostics and treatment. But the bill? Even in the raging 2022 inflation we're experiencing now, the bill was "only" $520. Large dogs are inherently more costly from a health standpoint.

Certainly, the dog health insurance industry knows this. According to *Forbes*, large dogs crowd the list of the most expensive breeds to insure: dalmatians, Bernese Mountain dog, mastiff, bulldog, Doberman pinscher, and pit bull. And small dogs top the list of the least expensive dogs to insure: Maltese, Peekapoo, schnoodle, Chihuahua, Pomeranian, and Yorkshire terrier.[112]

Looking at data from insurance companies, it's easy to see giant and large breed dogs can be forty to sixty percent more expensive to own than small dogs. In fact, we insure our two Havanese for about the same we'd have spent to insure one golden retriever. From hip dysplasia, to bloat, to certain aggressive cancers, large breed dogs are often simply more susceptible to life-threatening, expensive-to-treat issues requiring significant veterinary care.

To be sure, owning any dog—large or small—is costly. Even in a good year, when you're not dealing with any unusual expenses, you can still plan to spend close to $1,500 annually on vet visits, food, treats, boarding, vitamins, grooming, and toys. According to 2021 data collected by the American Pet Product Association,[113] annual costs include surgical vet visits ($458), routine vet visits ($242), food (about $287), treats ($81), boarding ($228), vitamins ($81), toys ($56) and grooming aids and brushes (approximately $47). And that's a highly conservative estimate. The food and toy expenditures seem especially low, given the rising popularity of pricier, personalized dog foods[114] and dog owners' increased spending on subscription toy boxes.[115]

So, does dog size make a measurable difference in the cost of ownership? I know it *seems* as though Phoebe (and now Scout) are much less expensive to own than our goldens were, simply because their food intake is so much less. As noted previously, the ASPCA confirms my gut feeling, revealing it costs much more annually to own a large breed than to own a small dog. Of course, it's also true that small dogs live, on average, five years longer and thus cost more over a lifetime. In my mind, though, it's only the annual expense that's relevant. The lifetime cost doesn't really count, given most dog lovers are never without a dog.

But again, that's not to say small dogs are inexpensive to own. The Yorkshire terrier and the Cavalier King Charles spaniel each made GoBankingRate's list of "30 Most Expensive Dog Breeds."[116] And if, like us, you own a breed with hair that grows, your grooming expenses will cut into any small dog savings. Indeed, your spending on haircuts can quickly set you back as much as your own salon appointments. However, because you *can* learn to do the job yourself, grooming is technically an optional expense.

In fact, it's the veterinary visits that create the biggest non-discretionary spending for dog owners. According to data gathered by the American Pet Product Association, of the $123 billion spent on pets in 2021, over $34 billion of that expenditure was on veterinary care and related products.[117] Nearly every source we consulted says the average emergency vet bill tends to fall in the $1,000-$3,000 range. We wish we'd been so lucky with our golden retrievers, whose bouts with bloat and cancers often meant our emergency bills easily danced near the $4,000 mark—

and that wasn't the annual cost. That was the per-emergency expense.

While we can't put a price tag on a family member's health, more than half of dog households can't afford these emergencies. According to the American Veterinary Medicine Association, in recent years most dog families earned less than $59,000 annually, and such households can't absorb extra expenses.[118] Thus, for good reason, cost matters in determining what dog will fit best into family life.

And the hard truth? Whether you adopt a large dog, giant dog, medium dog, small dog, or toy dog, they're *all* expensive. However, small dogs may be less prone to major injuries (though not if you're dealing with the spine issues that haunt many small breed dogs, including my courageous Chiweenie granddog Rosie), and they are much less expensive to feed. They're also often less costly to find sitters for and to travel with. If your discretionary funds are limited, then small dogs give renewed meaning to the term "purse puppies." They're literally and economically a much better fit.

CHAPTER NINE

SMALL DOGS LIVE LONGER

While small dogs take a tinier bite from our wallets, thankfully they'll take up a longer span in our lives. Small dogs generally live a very long time. And in all honesty, that's initially the only reason our family reluctantly made the move from beautiful golden retrievers to the pipsqueaks we have now. We just couldn't continue taking part in the cancer cycle embedded in the genetics of so many big breeds.

As we were raising our three children, all three of our golden retrievers died way too young from fast-moving cancers. And two of our beloved dogs died of the same aggressive, nearly untreatable cancer—hemangiosarcoma.

The first loss was Maddie. We had adopted her, our second golden, from a family who didn't want her anymore because she had a hip issue. At two years old, Maddie was a love whose only fault was occasionally sneaking through three backyards to visit an elderly neighbor who she must have sensed needed company as he puttered alone in his garage. At ten years old, when she suddenly began having difficulty getting up from her favorite spot, we took her in for testing.

It wasn't good. They diagnosed her with hemangiosarcoma of the spleen. We fought this enemy with both conventional and holistic Eastern medicine, neither of which did any good. Or maybe they did—maybe our futile efforts to lengthen her life at least made her more comfortable. But no intervention could save her. She died within months of the diagnosis.

We knew, of course, many hunting breeds—especially Labs and goldens—were living shorter lives, but we didn't let that stop us from bringing them into our lives. We just assumed *we'd* do better and beat the cancer. First, we resolved to find a line of goldens from a breeder committed to facing and fighting these life-shortening health problems.

And we found her. While we weren't yet ready to love a new pup, as we were still grieving Maddie, we happened upon an incredibly knowledgeable woman who followed the golden retriever cancer issue closely.

We developed a relationship in the ensuing months, and she agreed to let us co-own one of her amazing girls, Haly, whom we brought into our home at around nine months old. We committed to ridding her environment of carcinogens as much as possible. We researched and continued to feed the organic raw diet she'd been given since puppyhood. We exercised her, we kept regular vet appointments and then some. We loved her completely. It did no good. Haly died of a sneaky mast cell tumor at eight years old.

Her beautiful son Romeo died at eight as well, from a hemangiosarcoma like the one that killed Maddie. The silent tumor in his heart suddenly ruptured one night, and it broke us as well.

We knew then we just couldn't watch another one of our family members die of the cancers plaguing so many large breeds. After experiencing the crushing early loss of three wonderful retrievers, we resolved to find dogs who could stay with us for more than a scant decade. So small dogs with long life expectancies were where we turned. When it comes to big dog versus small dog lifespans, the comparison isn't even close. While life isn't guaranteed, at least our smallest family members have a fighting chance to be with us for a long time.

Recent research from the Dog Aging Project confirms what many past studies show: the smaller the dog, the longer the life.[119] While the exact mechanisms of that long life aren't agreed upon, new focus on dogs and their aging is uncovering much that's important to all of us. Thus, in addition to looking at the genes responsible for longer life in small dogs, researchers are delving into the *way* small dogs age. They're finding interesting differences between big and small dogs.

One recent study found that, as they become seniors, large breeds don't break down glucose correctly and their use of Coenzyme A, which aids the body in producing energy, becomes inefficient. At the same time, small breed dogs have bodies seeming to become almost more efficient at producing energy as they get older.[120] It's likely we'll learn a great deal more in the coming decade, as the concept of One Health[121] drives us to better understand the interplay between the well-being of every creature, every plant, and every ecosystem.

Dogs, especially, provide particular insight into human health, given their faithful partnership with us in nearly every aspect of their lives. They sit where we sit, they walk where we walk, they breathe what we breathe, and they often eat what we eat. While genetics may drive many aging differences between big dogs and small dogs, it's also likely more is at play. The environment, to be exact.

Perhaps because small dogs are less often on our floors and more often on our furniture, or they're infinitely easier to exercise, or they may eat higher quality foods, or probably receive more frequent baths—any and all these factors could contribute to lengthening their lifespans through better nutrition and lesser exposure to environmental toxins.

While these are just theories, the science is catching up and now that researchers are discovering *what* is different about small dogs and their aging, it's only a matter of time before they find out *why*. Those discoveries will improve the lives not just of all dogs but of all humans.

CHAPTER TEN

THE SMALL DOG LIFE FITS EVERY LIFE STAGE

While we're discussing longevity, let's talk about our human lifespan and one of the best things about small dogs: From a purely physical standpoint, small dogs are so much easier on everybody's body. Petite pups are portable and simply easier to control, even if untrained. They don't strain the back, they don't yank your shoulder socket when lunging at errant squirrels, and they don't require physical effort to wash in the sink. The only major danger they do pose is a tripping hazard—sometimes you don't see them underfoot. But that's nothing a little awareness and training can't cure. Whether we're nine or ninety years, most of us can handle a small dog. They fit beautifully into all our life stages.

Why is that? Well, to grasp the importance of small dogs in the world, we need to contemplate the vital role of dogs in general.

The Small Dog Life Fits Every Life Stage

As our understanding of animals deepens, there's growing recognition of our dogs' consciousness. And we're not crazy for seeing it. As we discussed in our chapter on small dog smarts, modernized scientific studies now test canine self-awareness using a species' appropriate model, rather than the mismatched human model used in the past. In addition, when human models are employed, they're suitably adapted. This fresh approach reveals dogs do, indeed, understand their own consciousness and enjoy self-awareness in much the same way we do.

Dogs recognize their "beingness" in a "I think, therefore I am" sort of way.[dd]

You'll remember the multiple studies we discussed confirming most dogs possess awareness equivalent to that of a human toddler. In the past, studies testing an animal's self-awareness used species-*inappropriate* tests like the mirror test, in which dogs were dismissed as lacking a sense of self because they didn't recognize themselves in the mirror. Tests like these are now widely acknowledged as unfair measures of dog cognition, because (a) dogs are not creatures who rely heavily on sight to navigate the world and (b) they're not so superficial as to care about their reflections.[122]

But why are we discussing general dog sentience in a chapter about small dogs and life stages? Let's turn to demographics for part of the answer. Fewer people are marrying today and are instead opting instead to live in single-person households. More couples are delaying children or deciding against them altogether. Couples with grown children are no longer overrun with grandchildren

---

[dd] What beautiful cosmic irony that Descartes' barbaric animal machine theory destroys itself with his own words.

143

within five or ten years of discovering their empty nest. And, thankfully, seniors are no longer doddering quietly off into the sunset to sit under quilts.

Do you see the pattern? Science may not have caught up completely, but most dog lovers understand the profound creature connection dogs bring. We know dogs understand their beingness, and such awareness is crucial to our deep friendships with these beings. Burgeoning dog ownership is occurring in part because, in many instances, people without children at home are now nurturing and enjoying companionship with another species entirely. And if they're not, it may in part be because they worry about the challenges dog guardianship brings. It's in these instances, especially, that small dogs make such perfect sense.

Market research firm Packaged Facts reports that, in the past decade, dog ownership among those who've never married has increased thirty-two percent and with lucky retired humans, that ownership has grown by a staggering sixty-seven percent.[123] Just as childless homes are contributing to this boom in dog guardianship, families are also expanding their canine numbers. So, before we look at the benefits to small dogs filling child-free homes, let's view the important ways small dogs provide enormous benefits to families with children.

The Case for Small Dogs and Kids

From the window of my bedroom upstairs, I stand silent witness to the bedlam below. It's not physical mayhem—not yet anyway—but more of a fever-pitched battle of wills.

Young, indignant wills.

The Small Dog Life Fits Every Life Stage

With my forehead to the windowpane, I stand wondering whether the seven kids bickering vehemently on the lawn below will work it out themselves this time.

Or whether, as usual, I'll have to intervene.

Three of the angry children are my own, joined by four of their equally outraged best friends from the neighborhood.

It's the traditional afternoon baseball game, complete with the even more traditional devolvement into bench-clearing bickering. The argument is invariably some variation of the same refrain involving fair play, or lack thereof: whether the ball went foul, whether the pitch was wide, whether so-and-so's foot missed the base, or whether it was really their turn to take my four-year-old daughter onto their team.

As always, their bases consist of our Japanese maple tree, a whispering poplar, and a mossy oak tree. And then there's home plate, whose identity varies for convenience's sake: Sometimes it's someone's shoe, sometimes a stained backpack.

And sometimes, it's the Buddha statue from my garden.

You'd think he'd halt the bickering himself, but no.

Nearly every day after school, they play. And nearly every day, within fifteen minutes of their start, they can't help themselves. An argument breaks out. And in response, I often must step in. Or at least I send my representative.

I shake my head, call our golden retriever Maddie, and we trot downstairs together. Opening the front door, I yell "Game's over!" as Maddie rushes past, knowing it's her job to find the baseball and run with it. Stealing the ball at their feet, she curves at top speed around the house, the loud, laughing protests from the kids trailing after her.

DOWNWARD SIZING DOG

It's the suburban dream depicted in movies, television, and SUV commercials: parents, two kids, and the golden retriever piling into the car for Saturday soccer. Or, those lush, green lawn ads, in which a gentle, giant Labrador lies with his stomach skyward, while toddler grandchildren giggle, throwing themselves in a pig pile on top of him.

It's a beautiful picture and swapping out the retrievers for a rat terrier admittedly doesn't elicit the same charm. But Shakespeare's and Instagram's opinions to the contrary, life really *isn't* a stage, and the happy moments with kids and big dogs exact a pretty price; as we've mentioned before, the cold truth is big dogs are big work, much more so than small dogs.

We've gone over the fact big dogs shed. A lot. Either the fluffy stuff gathers in tumbleweed conventions on every baseboard of your home, leaves a fine coating on your carpet and clings to socks, or it sticks, like porcupine needles, into your upholstery. I know pointing that out again might be repetitive. Okay, it is repetitive. But clean homes are even more important and less attainable when you have small kids. Adding lots of dog hair to the mix can just make life gross.

Kids drop things: toys, pacifiers, food. And then they pick them up and, depending on their age, put either the thing or their hands somewhere in or near their mouths. Ingesting dog hair won't kill them, but ugh.

Kids of all ages crawl around or lie on the floor, usually in their pajamas or joggers or yoga pants, all of which just love gathering lint and dog hair. If the dog hair isn't on the floor, it's on your big dog, who you probably have had no time to brush. Don't worry. Much of that dead hair will wind up on your kids' clothes, which you'll be happy to wash (again) in your spare time. And then you can clean the hair from the washer. The lint trap will grab the rest. Just remember to wipe it down every time.

Big dogs don't stop at covering your kids with their hair, either. Their waste is prolific, and some of it inevitably gets overlooked in the grass, and someone more inevitably steps in it and, even more inevitably, tracks it into the car, or worse, the house. And, because cosmic irony seems to work best in crazy-busy households with children, dogs, and packed schedules, such disgusting mishaps occur just when you're running late for the next soccer game or dentist appointment. Or worse, a playdate with that super-fastidious neighbor.

Of course, the goldens we were privileged to live our lives with—Maddie, Haly, and Romeo—brought so much joy into our home I didn't begrudge the daily vacuuming, or the thrice-weekly poop patrol. Well, sometimes I did. But I love huge dogs, and while growing up and into my mid-adulthood, I believed a dog was only a dog if it resembled a small horse. So, as soon as we had the space in our home and our yard, a big dog I insisted it be.

I didn't see small dogs as potential playmates for my children. Such creatures (the dogs, not the children) seemed fragile and not great at romping in the yard. And to be sure, certain small dog breeds (like many large dog breeds) are wholly unsuited to life with small kids. But that's not true of most small dogs, even though I believed it to be the case for most of my life. Like all prejudices, mine was based on faulty assumptions. The foolishness of those preconceived notions became even clearer to me after I spent some time speaking with a mom of five who's much smarter than I ever was about the issue.

"It's not just about having a dog that fits into a narrow slice of your life. It's about having a dog that fits in all parts of life. Small dogs bring families together. Big dogs, on the other hand, can be isolating." Diane[ee] has just finished telling me about her sister's German Shepherd, a strong, boisterous herding dog who tries to dominate and herd children at every turn. They just can't get him to play nice. As a result, the sister takes the dog for walks, alone. The sister feeds the dog, alone. The sister grooms the dog, alone.

The sister loves the dog.

But their situation resembles the parallel play we see toddlers share. She and the dog play. She and her kids play. But the two cohorts don't mix well. To help the shepherd become an authentic part of the family, Diane's sister will spend extra money and extra time training her dog to be something other than what he is in his core—a working dog designed to work.

In contrast, Diane's two Havanese are a shared family experience. The well-behaved pups even take orders from the youngest human in the family, who is all of four years old. As Diane puts it, there's so much to train out of a large dog. But a small dog makes much smaller and less damaging mistakes. They're easier to train, and they're easier for all family members to care for. Everyone pitches in to feed them, groom them, and play with them.

With training, the whole family takes part. With any dog, big or small, Diane points out it's as much about training the kids as it is training the dog. And so, all her children at one point attended puppy classes. And, unlike

---

[ee] Again, names and breeds have been changed for the sake of sibling harmony.

a seventy-plus pound young male shepherd, the Havanese are easily directed by even the smallest child.

As Diane watched her sister begin looking at large dogs to round out her family, her own life-experience with big dogs and small kids prompted her to suggest her sister find something smaller. Diane and her husband had owned large dogs, including cattle dogs. At one point, they adopted a second dog, a husky/shepherd mix who nearly tore their house apart wrestling with the cattle dog. So, Diane gently advised her sister to consider the small breed they'd come to adore—a Havanese. But no, her sister wanted a dog "who could really hike with us."

Fast forward two years, and you'll find Diane and her husband, their five kids, and two Havanese hiking several times a week. In the winter, they do a 2K loop after school. In summer, they all navigate the 5.6-mile hiking trail near their home in the Pacific northwest. The dogs trot along happily, and the whole family loves these times.

And Diane's sister? The one who adopted the shepherd—that "big dog who could hike"?

Her dog's not on these outings.

He's too much for them to handle, and so he stays at home, further building up energy.

In a family with young kids, that's the difference between big dog dreams and big dog reality. As Diane says, most big dogs were bred to work. They were only companions after their hard work was done and they were exhausted from the day. They don't easily fit into our modern lives where busy families with small kids are already scheduled beyond themselves. While exceptions exist, many families can't naturally accommodate a big dog.

But small dogs? They look up to our kids; they teach them empathy, leadership, and friendship. They mesh with all parts of our lives. And, in this time when mental health challenges are hitting our children particularly hard, small dogs may provide vital emotional support and coping mechanisms that aid them in navigating a childhood where the leading cause of death is gun violence, and their closest friend is a blue screen.

Even a brief search of the scientific literature on pets and children provides enough positive research to keep you reading for years. Since the 1980s, the topic has gained traction, but it's only in recent decades that dogs became a more targeted focus. Here, the research uncovering the ways dogs support our children is profound.

One recent, controlled study found that dogs in schools significantly reduce stress (and therefore improve learning) for elementary school students. What's particularly important about this study is the list of breeds involved, over one-third of which were under twenty pounds and included a Yorkie and a Pekinese.[124]

A 2022 comprehensive literature review reveals "unequivocal evidence" dogs lower stress in both students and teachers, and most significantly, improve both learning motivation and participation.[125] Another study tells us elementary teachers see their struggling readers engage more enthusiastically and show more confidence when reading to dogs.[126] Given the emerging theory that perhaps dogs contributed to the development of human language itself,[127] it makes sense our pups also help our children as they grapple to learn.

And it's not just in school where dogs infuse kids' minds and hearts with superpowers. Children with dogs generally experience less anxiety and loneliness and report higher self-esteem.[128] Kids on the autism spectrum bond with their family dogs in ways that may aid their social skills—and the kids with *small* dogs seem to have closer connections with people.[129] In fact, children with autism who have a service dog learn to understand facial expressions in a way children without pups do not.[130]

And for those of us with teenagers in the house, know an adolescent's close relationship with their dog could also improve family and friend relationships.[131] Researchers guess a teen's dog can ease tensions or act as a connection to others, spurring increased family time. Whatever the reason, dogs and teens are almost always a wonderful combination.

So, we know our kids' psyches improve around dogs, and in particular, at least one study proves some children prefer small dogs. But what about a dog's influence on our children's physical health? We've already learned of the anti-allergy benefits dogs may provide infants, but a pup's support of baby humans goes far beyond that. Groundbreaking research out of Japan found infants and toddlers raised with dogs were remarkably less likely to suffer developmental delays.[132] And children and teens who live with dogs are much more likely to get the exercise they need.[133]

Although few studies currently exist, I can tell you anecdotally that small dogs likely provide many more opportunities for exercise than large breeds. With a rambunctious large animal who's not perfectly trained, you risk having your eight-year-old pulled onto the pavement. In the yard, it's the dog getting most of the exercise from play. And inside the house? Would *you* encourage a seventy-pound dog to play chase?

In terms of the way dogs influence our physical activity overall, one study reported no true differences between big and small breeds. Thus, we once again see the flawed public perception that big dogs belong to active people and small dogs belong to those of us who sit on the porch. In addition, I'd like to point out that, of the nine breeds taking part in the study (Rottweilers, Parson Russell terriers, Jack Russell terriers, Westies, Bernese Mountain dogs, Cavalier King Charles spaniel, whippet, border collie, Labrador retriever, and Belgian shepherd dog), it was the small Westie who often placed in the top two with respect to getting their owners up and moving, coming in second only to Rotties.[134]

With small dogs, even the youngest child can be useful and in charge of the leash on walks and can do so safely. In the yard, small dogs often enjoy running *with* kids as opposed to playing fetch, inspiring running, leaping, and jumping—for the kids and not just the dog. And indoors, we often play impromptu games of chase or hide and seek with the dogs, which we could never risk with Romeo, who became a golden wrecking ball when wound up.

Small dogs and children are a beautiful team. Having said that…

There's no denying mixing small dogs and small kids requires extra care. Yet, such vigilance is just as vital when combining big dogs and toddlers. In neither case should they be left alone together, because it's all fun and games…(wait for it)…until someone gets hurt.

Some dogs, big or small, are not great around kids. Nearly every study published will state the obvious: Golden retrievers and Labs are very low on the bite risk scale. Our Romeo was a case in point. He would willingly endure Phoebe's onslaughts and herding, he would approach with gentle caution humans holding out food, and he instinctively and immediately laid down to toddler eye level, even when said toddler suddenly barreled at him at top speed. We often say the opposite of other breeds.

We all know the big breeds with bad reputations, and while we'd argue (and win) that those ill opinions are unfair, that's not the subject of this book. But what about the small dogs who might not readily tolerate being startled or loomed over by kids? While we don't think stereotyping breeds is at all wise, it's still important to understand the experience of others.

According to the American Veterinary Medical Association's Animal Welfare Division, certain small-to-medium breeds *do* show up more often in reports of snappishness toward people.[135] In particular, among small breeds, Shih tzus and Lhasa apsos appeared frequently in one Canada survey of owners.[136] However, just as I've said above, the AVMA cautions us all to look at the logic behind the data: The dogs aren't the issue. It's the people. Socialization and supervision matters.

Dogs with acute anxiety, dogs without deliberate and frequent exposure to children, or dogs left alone with toddlers are more likely to bite, whether out of fear or guarding behavior. In fact, according to the association responsible for accrediting veterinary hospitals, in many cases, dog bites occur "because the kid started it."[137]

And that's why, no matter the dog's size, we must always ensure that, while our children are benefiting, the dog is an equal and protected partner in all of this. Doing so is becoming a recognized responsibility of anyone engaged in animal-assisted therapy. It's no longer just about us. It's about our dogs as well. In fact, the global One Health philosophy argues any interaction between people and animals should benefit all—human and dog—equally.[138] Fortunately, growing recognition is leading to better education and ways of measuring "dog happiness," particularly for families with children.[139]

Approaching the melding of small dogs and kids from a "we're all in this together" standpoint can virtually guarantee the pup's well-being and happiness. As we're enriching the small dog's life, we're providing so much more to our children. From infancy, we're jump-starting their immune system and bolstering their development. We're encouraging their physical health and activity at rates simply not seen in dogless families. We're nurturing their empathy, problem-solving, and leadership skills. We're teaching them to care and to love.

What did I say earlier? Small dogs and kids are a beautiful team.

The Case for Small Dogs and Millennials

If you're a millennial, you represent the generation that gets it. You're the ones who coined the term "fur baby" and you proudly refer to yourselves as "pet parents."[140] You revere animals on a larger scale than your boomer parents (though, as a boomer, I'd match my animal awe with your animal awe any day), and you deserve praise for doing so. It's not surprising nearly three-quarters of millennials have pets.

In one survey, half of you say your household includes a dog, and you consider your dog as one of your children.[141] And we all know how parents sacrifice for their children. Thus, it's not shocking to learn nearly twenty percent of millennials have "gone without" in order instead to purchase something for their pet.[142]

But let's talk about the reasons millennials are so very close to their pups. In part, it may be, in 2019 anyway, a whopping forty-four percent of you weren't sure you wanted children. At least one market research company guesses your reluctance to bring kids into the world is tied to your somewhat precarious financial state, the well-known career-smothering challenges of motherhood, and, maybe, you just want to enjoy freedom a bit longer.[143]

If you're a millennial choosing dogs over children for these reasons, then the smartest option is a small dog. Financially, we've already seen that small dogs are generally less costly to maintain, so you can reap the benefits of canine company and still pay down your debt. Or, if debt isn't an issue, but you're currently building your professional life, from a career and time perspective, small dogs' portability and smaller ecological footprint make them the ideal way to round out your family. And a whole lot easier to talk your employer into letting you bring your pup to work. (Just ask Brooke from Chapter Five.) Finally, when and if you *do* have kids, we've seen why perhaps a small dog makes the best first child.

The Case for Small Dogs and Boomers

While millennials deserve praise for raising the bar on how we care for the dogs who share our lives, boomers deserve credit for opening America's eyes in the first place. As we explored in early pages, our relationship with dogs began for purely practical reasons: We needed them to guard our flocks, herd our cattle, and drive off vermin. But even after most families shed such requirements and kept dogs merely for family companionship, pre-boomer society dictated *dogs* belonged in *dog* houses.

And those dog houses belonged in the backyard.

Away from the house.

During child-rearing decades for the vast majority of the Greatest Generation[ff] and the Silent Generation,[gg] it was unheard of to let dogs in the home, let alone on the bed.

Along came my generation, the boomers, and we pulled our dogs into our cultural revolutions…and into the house. We were the first generation to make our dogs true family members. We brought dogs inside our homes, our hearths, and our hearts. Now, the youngest boomers are coming in hot into our early sixties. Unlike our parents' generation, who generally washed their hands of pets as the children left home, we modern empty nesters keep our dogs longer *and* adopt puppies in numbers not seen in prior generations.[144] And for very good reason. The relationship boomers enjoy with our dogs is transformative, both mentally and physically.

Let's look at the life-changing ways dogs improve boomers' physical health. We'll start with the heart, as aside from recent statistics accounting for COVID-19, disease in that organ is usually listed as the number one killer of people over sixty.[145] But a boomer's risk of death from heart-related causes is much, much lower if we own a dog. Study after study proves having a dog not only makes us less likely to have a heart attack, but it also lowers our risk of death if we do.[146]

[ff] Americans who came of age around World War I

[gg] Americans born between 1925 and 1945

The Small Dog Life Fits Every Life Stage

Mayo Clinic recently took a deep dive into studying the link between dog ownership and cardiovascular health, naming the following seven keys to a fit heart: body mass index, healthy diet, physical activity, smoking status, blood pressure, blood glucose, and total cholesterol. By combining these factors, researchers determined a score predicting one's risk for cardiovascular disease. The verdict?

Dog owners possess much higher heart health scores than non-dog owners.[147]

Similar results were shown in a recent comprehensive review of all studies connecting human longevity to dog ownership. Here, study after study reveals having a dog significantly reduces mortality rates—*by twenty-four percent* in fact.[148] The benefits of a dog may be most profound for people living alone; here, dog ownership lowers mortality dramatically.[149] Given that well over a quarter of the sixty-plus crowd in America lives alone,[150] finding canine companionship to ease isolation seems like an easy decision.

Debra, a retired mortgage underwriter, provides perfect proof small dogs are virtually essential for a life well-lived. Like many of us, Debra didn't start out loving small dogs. When she was raising her three boys, their dog, Sundance, filled Debra's family with golden retriever love—and hair. After Sundance passed at fifteen years of age, her former husband brought another golden retriever, Samantha, to his new home. "Sam" would often tag along with Debra's teenage son, traveling with him between mom and dad's house, so Debra's big dog connection continued.

Then, when Sam died and her son eventually left home, Debra found herself dogless for a long while. Living alone with a demanding work schedule, she believed it would be unfair to keep a dog. But then she began dating a veterinarian, who introduced her to small dog bliss in the form of a bichon named Chester.

Debra fell in love—with Chester, not the veterinarian (though she may have loved him too; I didn't believe it polite to ask.) And along with Chester came the Havanese belonging to the veterinarian's daughter. Debra fell even further in love with this dog, his pint-sized affection, and his quirky sense of humor. Now immersed in the small dog camp, she jumped at the chance to adopt a dog when her company offered remote work about six years ago. She brought home a three-month-old Havanese puppy—little four-pound Oliver. Now, at only six years old, Oliver has accompanied Debra through some watershed life passages, including retirement and a major move. He's become her essential companion.

In listening to Debra recount life with Oliver, it's striking to hear how deeply Oliver influences her good health. Though she has a fenced backyard, she and Oliver venture out in all kinds of weather four times a day, every day, and they walk about a half-mile each time. She confides "even when I don't feel like it, he loves the walk, and it gets me out." There's a gate at the back of Debra's land backing up to a walking trail where they encounter deer, bunnies, and squirrels. And they always meet other dogs walking their humans. She laughs when she confesses she and Oliver remember the dogs' names but not the names of their people. Either way, they're both achieving vital exercise and making connections that protect the heart.

In addition to aiding boomers' internal health, dogs ward off external hazards as well, alerting us to environmental dangers we might miss as we grow older, including house fires or things that go bump in the night.

The Small Dog Life Fits Every Life Stage

Our sharp senses dull a bit, and our sense of smell and hearing can miss early warning signs of potential disaster. As we saw in Chapter Three, our dogs quite simply keep us safer.

So, for boomers, won't any dog size do? As a demographic, baby boomers amassed significant disposable income,[151] and with well-established careers and grown children who seem to be holding back on making them grandparents, they revel in more free time. Boomer status likely makes us more financially and emotionally able to adopt a large dog.

So why do small dogs still make more sense?

For one, there's the portability factor we've talked about. With extra leisure time, many of us Boomers plan four to five getaway trips annually.[152] (And yes, I'm conveniently ignoring the pandemic that's plagued our world. As we learn to live with new disease patterns that aren't going away, we're returning to travel, and indeed, AARP reports we're eager to get back out there.[153]) As we covered in Chapter Seven, small dogs travel well. Big dogs usually need to stay home.

And while my boomer generation doesn't want to leave our fur-kids behind when we travel, we often *do* want to leave behind the big house and large yard. Permanently. Much like our millennial children living in tight square-footage, many boomers want smaller homes and/or condos, which call for smaller dogs as well. Indeed, sometimes they demand it: Many homeowners' associations place size restrictions on dogs allowed on the premises. As we plan for possible down-sizing, we'll enjoy boomers' better living options if our pups are twenty pounds or less.

That lighter poundage brings bonus benefits in the form of a family member that's easier on aging human backs, shoulders, and joints. Lifting the pups into and out of cars or helping them up the stairs or onto furniture just won't be an issue if our pups are small. In addition, with small dogs, we're much more likely to be walking them instead of them walking us.

And as we've pointed out previously, small dogs are likewise easier on aging pocketbooks, as boomers approach retirement and stare down the barrel of a fairly fixed income. Small dogs possess all the benefits of canine companionship, minus many of the drawbacks. Those benefits increase as we age.

The Case for Small Dogs and Pet Parents in Late Seventies and Beyond

From a physical standpoint, small dogs are so much easier on our human bodies, especially when we're aging. Even a well-trained large dog needs help getting into the car or must be handled in the bathtub or lifted in case of an emergency. Sometimes, one person just can't handle a large breed. And for most of us, that difficulty becomes exponentially more challenging as we age.

We've experienced this ourselves, the night Romeo died. Despite believing we'd escaped hemangiosarcoma, the cancer that took our Maddie with such brutal efficiency, that was not to be. Just months before, Romeo had received a clean bill of health—at least as far as his spleen and blood work went. But hiding in his heart was a tumor no one saw. The evening Romeo fell ill, he collapsed and, although conscious, he could not stand. It took three adults—me, Ron, and our daughter Caroline—to carry his limp form to the car on a stretcher we cobbled together from blankets. Our quick work meant that in a matter of twenty minutes, Romeo was at the emergency vet hospital.

At the time, Ron and I were only in our mid-fifties and extremely active and healthy. If the three of us had trouble in such an emergency, someone living alone, no matter how young and in shape, probably couldn't have lifted Romeo's barely conscious bulk at all. And at eleven o'clock at night, when time is of the essence, that's a problem. But with a small dog? Almost anyone can pick them up and get them where they need to be.

Of course, portability in a crisis is hardly the sole reason to select a canine companion. But as we age, it's certainly a huge consideration.

All the topics we've discussed in previous chapters—the mind and body benefits of small dogs—hold perhaps truer when we're seniors. All dogs are a triple threat against aging: They heal us physically, emotionally, and even spiritually, keeping us connected both to the present and to something greater than ourselves. To be sure, any pet, even the non-cuddly kind (I'm thinking tarantula), can provide anti-aging benefits. But dogs have proven themselves to be our helpmates since time immemorial. The quality-of-life improvements dogs offer us reverberate most strongly with those of us in our late seventies and beyond.

Indeed, even that taskmistress known as Science, in a rare show of emotion, chastises those who "underestimate or belittle" the importance of animal allies, pointing out pets stave off isolation and inactivity for the elderly.[154] A recent case study involving Violet, a seventy-seven-year-old woman and her miniature schnauzer, Jack, puts it in personal terms: Jack provides daily companionship and a reason to engage in life for a woman who's seen her share of challenges.[155] Health issues resulted in Violet undergoing multiple amputations, and financial hardship has her living in a one-bedroom subsidized apartment. She uses a prosthetic leg in the apartment and her wheelchair when she's out.

Yet as this study revealed, neither Violet's health nor her perception of her financial situation suffers because of Jack. In fact, he connects her to a full life through his presence and the need to care for him. She talks with him, informing him of what's coming next in their day. He provides physical affection, and he naps next to her. He gets them both out of the house every day, as she walks him. If one small dog can do so much for an elderly woman who faces such challenges, envision the effect her improved sense of purpose and self must have on others around her. And then imagine if more of us kept and cared for small dogs as we navigate toward our one-hundred-year mark.

Researchers in Japan did just that, following the lives of nearly twelve thousand Japanese residents aged sixty-five to eighty-five for three years, unfortunately having to cut the study short because of the onset of COVID-19. Their most important finding? People who owned dogs were half as likely to have suffered a disability during the study period than participants who'd never owned dogs.[156]

Now, let's take an elderly man—we'll call him "Richard"—whom I've cobbled together from several humans I know. Richard is eighty-eight and lost his wife to a heart attack three years ago. His two children each live about an hour away and they visit him frequently, but his daily routine is becoming increasingly isolated. Back issues prevent him from golfing anymore, and most of his friends have died or moved to Arizona. Although Richard still lives in the neighborhood where the kids grew up, and he enjoys taking care of his yard and doing small repair jobs around the house, the neighbors with young children are at work all day. There are days when Richard literally has no human interaction.

So, Richard did what too many isolated seniors do—he took to watching television as a form of companionship. Sometimes, he'd sit down after breakfast and stay put until after the noon news. This extended inertia made Richard's back hurt more, and it was difficult for him to escape the chair. He joked the chair enjoyed him as much as he enjoyed it, because it just didn't want to give him up.

How would a dog change Richard's life?

Physical Benefits for Older Dog Owners

As we've seen in past chapters, having a dog benefits us physically. Whether it's because people who own dogs are more likely to enjoy exercise or because moving becomes more necessary when you own dogs, study after study proves dog owners exercise much more than do people in dogless households. Indeed, one recent study revealed dog owners are four times more likely to get the exercise they need than non-dog owners.[157] Even those of us with smaller dogs who need only short walks benefit tremendously. Having to rise from sitting to let your pup outdoors frequently can stave off the effects of aging.

After adopting Rex, a small terrier mix from the local shelter, Richard began walking Rex around the neighborhood after breakfast, and then again before lunch. Suddenly, he wasn't so addicted to the television's chatter. The walking made him feel better, and he added a third walk after dinner. All three outings were short—no more than fifteen minutes each. Yet that fifteen minutes, three times a day, added up to over five hours of exercise a week, enough to provide substantial health benefits and improve Richard's back pain and mobility.

Multiple studies confirm the experience of single, elderly people like Richard. We feel a responsibility to our dogs to exercise them, and when we're elderly and walk our dogs, we gain not only a boost in our feelings as caregivers, but a marked improvement in our physical health. It makes us fitter, better able to navigate the day, and less likely to find ourselves at doctor's offices.[158] So, dog walking for the win in battling aging.

CONNECTION TO NATURE AND SEASONAL RHYTHMS

After adopting Rex, not only did Richard's health reflect all the benefits of exercise, but the simple fact of being outdoors, in rain or shine, made him happy. Whether a bird song prompted him to smile, or the scolding Sumo squirrel inspired a chuckle, these outdoor forays boosted Richard's mood...and that happiness stayed with him throughout the day.

Science agrees: Being outdoors eases depression and anxiety.[159] And, nature's healing is particularly effective for the elderly. Research reviewed in a recent study of urban green spaces makes clear spending time outside—even in small doses—can improve our quality of life as we age.[160]

As we saw with Richard, whose world opened to nature when he began walking Rex, small dogs can help us find good reasons to spend time outdoors as we age.

SOCIAL CONNECTIONS

Perhaps the most crucial role dogs play for us as senior owners isn't in connecting us to health or nature but to humanity itself. Anecdotally, we all know how our dogs practically require we chat with other humans we encounter on a walk, and any dog guardian will tell you dogs don't just help us connect with *them*; they help us build relationships with the people around us.

And science backs us up. In an article published in *PloS One*, researchers confirmed these findings, noting that from Perth to San Diego to Nashville, dog owners were not merely more likely to know their neighbors, they were more inclined to count them as friends who provide emotional support.[161] And, more recent research proves as we get older, dog walking enhances our neighborhood social life by nearly doubling our connections.[162] How do such findings play out with our fictional friend, Richard?

Now that he's outside in his neighborhood walking Rex three times a day, Richard is experiencing human bonds he didn't realize he'd missed. He knows the mail carriers' names—even the Saturday delivery driver—and they often stop to give Rex a cookie. His new neighbors and their kids run out to meet Rex and now the parents regularly stop to check on Richard and ask about his day. Another neighbor he hadn't spoken to in years invited him and Rex to sit on the deck and visit. These social niceties may be brief, but they've improved Richard's daily experience.

We know isolation, especially as we lose loved ones and connections as we age, can lead to increased disease, and shorten our lives. With their reputations for easing and even *creating* our social interactions, dogs help cure loneliness. They're often our confidantes, and even when we're one of those cranky elderly people who claim we're just fine all by ourselves, dogs can be our "purpose," giving us a reason for keeping a schedule and getting up in the mornings.

That's part of what Jack did for Violet in the case study we talked about earlier. Dogs are attuned to our humanity and to a significant extent can understand and respond to our emotions.[163] As we relate next, Phoebe and Scout are living testaments to the way dogs improve life for those of us experiencing cognitive decline. Clearly, especially if we're an elderly person living alone, the canine-human bond can be vital.

Illness, Aging, and Small Dogs

Another direct physical benefit to owning small dogs into our later years comes in their ability to help battle our ailments. Elaine and Michael, whom we met toward the beginning of this book, showcase this beautifully.

You'll remember they'd owned working dogs for years, but their last one, Jasper[hh], died in 2020. Three days later,

---

[hh] Again, names have been changed to protect privacy.

The Small Dog Life Fits Every Life Stage

Elaine was diagnosed with breast cancer. Losing Jasper was not merely a crushing loss. It also felt like an ill omen for Elaine's cancer. For Elaine, facing a health crisis in a dogless home was unthinkable.

Enter Javier the Havanese, accompanied by Elaine's beautiful insistence on embracing life, cancer-be-damned. Puppyhood is challenging in the best of circumstances, but voluntarily combining puppyhood with a health crisis in one's eighth decade seems, well, either very brave or very imprudent.

But it was far from that.

Javi became Michael's "farm dog" companion and Elaine's ever-eager confidante. His sensitivity to Elaine's physical health prompted him to cuddle with her often, providing an unconditional, healing presence only dogs can give. And, when Elaine had a break from treatments, Javi's puppy ways kept them busy and moving in ways they might neglect otherwise.

In addition to providing unwavering acceptance and emotional cheerleading for an aging population's physical condition, dogs benefit those of us who will face Alzheimer's or other dementia. Recent scientific literature reviews reveal that dog therapy improves our psychological well-being when we face any type of cognitive decline.[164] Science isn't sure why that happens yet (they're looking into it), but in the meantime, I think we lay people can give them some direction.

Without well-controlled studies, can we say how dogs, and small dogs in particular, are wonderful for dementia patients? At the risk of shocking the scientific community, I think we can. And I can prove it.

In our own family, we have Sarah, a beloved, accomplished woman in her eighties who's been diagnosed with cognitive decline and probable Alzheimer's. As the years progress, this vibrant human loses pieces of herself: the ability to identify birds she spent a lifetime studying; a funny family memory she now confuses and swears never happened; a moment in history she misplaces decades later than it actually happened.

And then in other moments, Sarah will recall everything with perfect clarity. But even those moments are bittersweet, as she can't trust herself. We can see it on her face, that confusion and realization that her brain plays tricks on her. And yet, despite this illness toying with her memories, personality, and sense of self, she is still very much a warm, dignified human being.

But that doesn't make it easy. For caregivers, it becomes increasingly hard to maintain constant patience and loving attitudes, especially in those moments when our loved one becomes angry with our quest to keep them safe. And on the other side, Alzheimer's and dementia patients obviously sense their caregivers' occasional overwhelming frustration and discomfort. The caregiver/dementia patient relationship has got to be one of life's most challenging. It is here where we see dogs playing a beautiful bonding role.

Just as dogs ease interactions between strangers, they do so among family as well. Thus, for example, when I was most exasperated with our Sarah's angry insistence it was safe for her to travel to an unfamiliar city by herself, it was Phoebe and Scout who intervened. We could sit, each of us with a dog on our lap, and speak rationally. Something about Scout looking up at her with concern, and licking her face quite seriously, eased her desire to leave. The argument was over as soon as it began, and all because of

The Small Dog Life Fits Every Life Stage

two creatures who can't talk and weigh less than twenty-five pounds combined.

The pups also aided us when life required our attention, and we couldn't just sit and spend time with Sarah. While our visiting nurses and even other family members tired of hearing stories we'd already heard three times in as many hours, the dogs never did. They'd hop up in her lap and she'd talk to them. They'd listen with rapt attention. Sometimes, I'd mistakenly think she was talking to me, and she'd respond happily, "I'm talking to my puppies." And I'd silently thank them for the help.

But it wasn't merely breathing space they gave me.

No. They modeled for me how to be a better caregiver. They showed me Sarah's happiness grew exponentially when someone heard her and did not argue or talk back. She didn't really want to take off to another city. She wanted to feel valued. She wanted to feel like the adult she is and not be treated like a child. Phoebe and Scout showed me how that could be done. After watching the calming, esteem-boosting effect of my pups, I didn't contradict or argue with Sarah. I'd remember to say "wow, that would be fun," or "huh, I didn't realize that," even when every ounce of my being screamed "I have to correct her."

No, I didn't.

The pups showed me that.

My experience with dogs and dementia isn't an isolated one. Memory care centers throughout the nation showcase the successes of their resident dog therapists. In addition, according to one senior housing organization, many nursing homes and assisted-living facilities allow pets, though they may place size restrictions. Here, again, our small dogs win, because such restrictions usually require the pet weigh less than fifteen or twenty pounds.[165]

DOWNWARD SIZING DOG

BringFido.com manages an updated list of dog-friendly assisted living and memory care facilities, many of which boast their own resident dog.[166] Of the sixteen communities featured, all but one shared pictures of residents with small breed dogs. Clearly, those who love and/or work with dementia patients certainly believe in the power of paws.

Which Dogs Are Right for Seniors?

Hands-down, small dogs are the best option for most seniors because toy breeds are generally physically and financially easier to handle. One veterinarian of forty years argues that, as far as big dogs go, only retired racing greyhounds would be appropriate for senior owners. In terms of small dogs, Dr. Mark recommends Lhasa apsos, Shih tzus, Yorkies, Chihuahuas, or Maltese.[167] These breeds commonly pop up in searches of "best dogs for seniors," in addition to Havanese (our personal favorite), pugs, and Boston terriers.

Occasionally, you'll see recommendations that seniors consider adult Labs or golden retrievers, but we just don't see those options as wise. As we've shared from our personal experience, in an emergency, they're incredibly difficult to lift or carry. Plus, they're hunting dogs with hunting instincts. They can pull you over in an instant. Even a twenty-pound Cavapoo can jerk someone's arm, as we witnessed today when Phoebe, Scout, and I greeted a seventy-something woman and her adolescent pup. The eager mixed breed dragged her owner across the road, tugging at her shoulder with surprising force. A small dog under fifteen pounds won't have the strength to accidentally hurt an elderly owner. Thus, unless a bigger dog is trained for therapy or assistance, it's just not something to risk.

The Small Dog Life Fits Every Life Stage

Perhaps most importantly, it's not fair to the dog if you're unable to accommodate its considerable exercise requirements. While smaller dogs can keep up with active people, they don't necessarily NEED the same amount of activity as larger dogs—even senior retrievers stay active and need significant exercise. Small dogs, on the other hand, especially those specifically bred as companions, provide all the benefits of dog ownership but in a pint-sized package that better fits the financial, time, social, emotional, and physical aspects of life.

PETS AND ELDERLY: SPECIAL CONSIDERATIONS AND SOLUTIONS

One major drawback to having small dogs when we're seniors is the dog's increased tendency to become a trip hazard, especially in puppyhood. And for senior humans with aging bones, that tripping can lead to serious, life-shortening fractures. Back in 2009, the CDC studied nonfatal falls attributed to dogs and cats. Of the approximately 86,600 falls annually requiring emergency room visits, eighty-eight percent involved dogs.[168]

While one would think most injuries would occur in older populations, in fact injuries were highest in children up to age fourteen and adults aged thirty-five to fifty-four. *Not* surprisingly, however, the highest fracture rates occurred in those over seventy-five years of age, and those are the injuries sending us to nursing homes or worse when we're that old.[169]

Diane, whom we met previously, is the poster child for trip-and-fall injuries among younger adults. When I first spoke with Diane via Zoom, she sported both a huge bruise on her face and a concussion. The cause? Her good friend's Havanese, "Dash," who exhibits conflicting emotions of (1) fear and (2) indignant outrage at passing cars. As a result, Dash invariably tangles humans in the leash as she retreats from, and then lunges at, four thousand pounds of steel, informing the vehicle she's the one in charge.

On one particular walk, even though Diane knew what to expect as a car passed, she couldn't avoid the fall when Dash darted left in front of her. Diane's face hit the pavement first, and she's lucky she escaped with only a concussion, bad bruising, and a sore shoulder. Because she's so young and healthy, her body absorbed the blow with relatively little damage. Someone in their eighties might have cracked an eye socket and broken a hip.

Like anything related to living with dogs, however, awareness and training of both human and canine is generally the answer. Of course, in Diane's case, she couldn't exactly train her friend's dog. In terms of awareness, if we're senior owners, we need to understand the very real hazards presented by having underfoot a small creature who follows your every move.

Many small dogs are "Velcro dogs," who will not let you out of sight when you're home with them. Deliberate mindfulness when moving in and around the house is good practice always, but especially so when there's another being in the house at ankle level. Using wide gates or other containment systems around the house can be incredibly helpful both for reminding us to watch out if we're elderly pet parents, and to prevent our precious pooch from bolting anywhere.

Another key inside the house is picking up toys and keeping bowls and beds out from underfoot. We keep a toy box in every room where we spend time with the dogs, and if you're smarter than we are, you'll train your pups to pick up after themselves. Although we've not yet managed that figurative and literal neat trick, it's apparently easy for them to learn. So do as I say and not as I do.

The CDC reports that trips and falls out-of-doors are reportedly caused most frequently by walking the dog: either the walker trips over the dog or the dog pulls or pushes the owner.[170] In each of these three cases, however, the owner winds up on the ground, injured enough to require an emergency room visit. Obviously, a small dog will not be big enough to yank or knock an adult human to the ground. But tripping is another story, as Diane's cautionary tale shows us. For this reason, it's vital that when we're seniors, we train our pups to be both comfortable and reliable with loose-leash walking.

While the CDC's record review of emergency room visits didn't categorize the age or type of dog instigating these trip and fall injuries, we're surely justified in imagining more of them occurred with high energy animals—either working or sporting breeds, or puppies. Anticipating and avoiding such animals is an easy workaround, as when we're seniors we can be purposeful about selecting calmer, companion breeds. As we've seen from veterinary and aging experts, those dogs are overwhelmingly on the very small side—under fifteen or twenty pounds.

An additional special consideration with respect to seniors owning dogs is the owner's fear about what happens if their dog outlives them. So many of us worry that, when we're elderly, we might not have the right to get another dog if the dog's life expectancy exceeds our own. And yet, how tragic and unnecessary is that conclusion?

Given any of us could go at any given moment, why should seniors forego enriching relationships with dogs because they worry they won't outlive them? In the same way we provide for our families, friends, and other important relationships through insurance and estate planning, we can address our dogs' well-being. Since the protection and happiness of our dogs is of utmost importance to all of us at all life stages, we discuss emergency planning for small dogs below.

The minor drawbacks to small dog ownership as we become what society labels as "elderly" are certainly easily addressed, and many forward-thinking communities are sprouting supports for seniors with dogs, from wellness check-ins to pup-sharing services. The major advantages to small dogs for seniors cannot be duplicated, even by human companionship. When one considers small dogs shower us with increased exercise, easier social connections, better moods, less loneliness, and unwavering, nonjudgmental emotional support, there's no question nearly every human—elderly or not—ought to have a small dog.

At Every Age: Emergency and Estate Planning for Small Dogs

No book about downward sizing one's dog could be complete without a section regarding the special life circumstances accompanying these tiny companions; namely, their wonderfully long lives. In fact, we can count on some small dog breeds on to live eighteen or more years, a number coinciding with legal adulthood in humans.

Just as we provide for the guardianship of our human children as soon as they enter our world, we should provide for our long-lived small dogs. It's something most of us don't think about. But we all need coherent plans that will kick in to care for our small dogs if something happens to us. And it was Instagram that spurred me to research these issues.

The Small Dog Life Fits Every Life Stage

I was scrolling Instagram several years ago and stopped at a popular dog account belonging to a fairly young woman. "Sophie"[ii] and Max, her Maltese, contributed daily to what is best about IG: Their feed was positive, funny, and ethical. So I was surprised when one day Sophie's post lacked its typical bright photo and airy text. But when I read the first line, it was clear why: She had just learned she had only months to live.

Knowing she had no family to take her beautiful companion Max—a small white dog who had a long life ahead of him—Sophie turned to social media to find him the perfect adoptive parents. She found those people, and now that Sophie has passed, her pup is cherished by other amazing humans.

Annually, thousands of people on Instagram remember Sophie's incredible dedication to her smallest family member. She sifted through hundreds and hundreds of potential adoptive dog parents, even as she was grappling with her own terminal illness. And she didn't cease until she found someone who would love and care for Max, just as she had.

As I read this year's IG posts about Sophie and her pup, who is living a safe, happy life with the guardians Sophie hand selected, I wondered how many of us fail to provide for our own pups in our estate plans. My husband Ron is an attorney; I am a former estate planning attorney, and yet our documents didn't provide for the love and care of Phoebe and Scout, who are cherished family members. Like most people, we assume our family will do what's best, and

---

[ii] Sophie's name and the type and number of her dogs has been changed.

yet that's not fair to the people we leave behind, and certainly not what's right for Phoebe and Scout.

So, what is the right thing?

You can go two directions here: the inexpensive, laid-back, cross-your-fingers nonbinding agreement, or the spend-a-small-fortune-hire-a-lawyer route. We'll cover both roads. And let's start with the path most of us will probably take, especially if our family relationships are incredibly stable, with great communication and trust among everyone.

If you enjoy such picture-perfect relations, maybe it's enough to leave written instructions regarding your wishes for your dog's care if you're no longer here or otherwise unable to provide for your pup in an emergency. Obviously, this path is less expensive but gives you (and the small dog you leave behind) no guarantees at all. Here's why:

> If you've promised your pup to a family friend, but your negligent heir wants her instead, your family friend will lose unless an actual contract exists. Binding contracts usually require lawyers.
>
> Technically, under most state laws, family pets qualify as personal property. So your dog will pass as personal property to whomever you named in your estate documents. If you've left no documents, state law determines who inherits your "property." So, it won't matter what you've said you want…if someone wants to make an issue of it, your wishes won't mean anything.

For guarantees, you'll need legally binding documents: a trust or a will.

The Small Dog Life Fits Every Life Stage

Not that many years ago, you'd be deemed slightly off-balance for even suggesting your estate documents should include provisions for your dog. If you lived in America, that is. In contrast, courts in England have long upheld wills or trusts designating funds to care for a pet. With few exceptions, however, for much of the twentieth century, most state courts ruled animals could not be named beneficiaries of a U.S. will or trust.[171]

In the 1990s, though, sanity prevailed when the National Conference of Commissioners on Uniform State Laws revised the Uniform Probate Code to specifically permit an animal to be named a trust beneficiary.[172] Most states adopted that code quickly, and, in 2016, Minnesota became the last hold-out state in the nation to grant these survivor protections to dogs. As of this writing, every state in the union and the District of Columbia allows pet owners to make will and trust provisions for their beloved animals.[173]

So now, no *legal* barrier exists to ensuring our pups are cared for if the worst occurs. Thus, any failure on our part to plan for our pets is not the result of a legal barrier. It's an emotional one. It's something I encountered often with estate planning clients: They'd meet, they'd make some decisions, and then their draft documents would sit in the bottom of their desk drawers, unsigned and unenforceable.

My theory? It's all wrapped up in the way many of us employ magical thinking about our untimely (or even timely) disability or demise. We, at some level, irrationally believe the universe is going to make an exception in our case, and unlike everyone else, we're living forever. Or we fear if we're prepared with actual documents, the end will surely follow quickly. But given our small dogs' long life spans, and their trust in us to keep them safe, we need to grow up and get the documents done. Perhaps the easiest place to begin is with a durable pet health care power of attorney.

A *pet* health care power of attorney is very similar to a human health care power, in that it names someone else to make healthcare decisions for your dog on your behalf. It's an incredibly handy document not only for dire emergencies but also for your dog walker and pet sitter if you can't be reached immediately. You can likely create this document without the help of a lawyer (though, I'm not recommending it—I'm just saying many people do).

Powers of attorney help in the event you can't be there to make health care decisions, but they would not survive your death and they don't allow you to provide funds to care for your dog. For that, you could create a lifetime trust enabling you to provide assets to be used for your pup's care both during your lifetime and at death.

Whether and how such trusts are enforceable after death depends on your state laws, which you can research yourself with the resources at Michigan State University's Animal Legal and Historical Center.[174] Use their interactive map to drill down on your state's statutes regarding trusts for animals.

At this stage, you may be thinking, "I wouldn't bother with a trust. I'll just put it in my will the next time I see my lawyer." But *is* your will the answer for providing for your dog after your death? Well, experts will give you varying answers, as experts tend to do. At the risk of sounding like the estate planning attorney I used to be, I personally would never use a will to ensure my pups' well-being.

Why? It takes at least a month, and possibly longer, to activate the provisions in your will through your state's probate court system. The language in your will, and the person and funds you designate in your will to care for the pup, can't and don't operate until the court approves your document. What happens to your pup in the meantime?

There is one solution if you *do* decide to work with your will. The New York State Bar recommends providing interim instructions that, hopefully, will operate to protect your pup.[175] (The resource cited is excellent, and provides sample language you might bring to your attorney, if they've not already created such provisions). Simply by discussing advance arrangements, securing the caretaker's agreement, and alerting everyone to that arrangement, your pup should be protected while the will is probated. But again, it's not guaranteed.

So, what's the best option? Stand-alone trusts, unlike wills, are already active and require no court action. Thus, they ensure your fur-family receives immediate protection (and love), and these trusts are legally binding. Pet trusts will set aside funds for your small pup's lifetime and provide clear, legally binding instructions regarding where she lives and who cares for her. These trusts can either be activated under your will (in which case, again, they won't operate until a court verifies that will) or in a trust you create during your lifetime.

That trust can be part of the living trust you use for your overall estate planning. I love the New York Bar Association's suggestion that owners leave a list of possible new owner candidates for their trustee to choose from.[176] Doing so lets your trustee make the best decision based on whatever circumstances exist at the time. So, for example, we might direct our trustee to choose first from among our children and then if none of our kids can take over our pups, the trustee could select from among several trusted friends. You can be as general or as detailed as you like, but you will want to consult your estate planning attorney to ensure the best chance of life running smoothly after death.

Enough about death, though.

Let's return to life, that thing we all want to keep experiencing. One important provision you need *during* your lifetime is an emergency card in your wallet, a durable pet care power of attorney permitting others to obtain health care for your pup if you're unable to do so, and notification to neighbors regarding what to do in an emergency.

One final word before we close the curtain on this depressing but necessary chapter on how to plan for your pet's future without you in it: Remember, I am not a lawyer anymore, and I don't play one on TV (or in this book or on Instagram or Facebook). The information above is to introduce you to concepts you might not have otherwise considered. It is not legal advice, and if you're not comfortable with unenforceable agreements you *must* consult your own lawyer regarding the individual laws in your state.

CHAPTER ELEVEN

BUT WAIT, WHAT ABOUT SMALL DOG SYNDROME?

But speaking of untimely demises, in my obsessive research for *Downward Sizing Dog*, I stumbled upon one particularly vitriolic article called "Eradicate Small Dogs Now."[177] In his essay for *Gawker*, writer Ken Layne variously labels a toy breed dog as "the very worst pet," "complete menace," "cruel joke," "freakishly shrunken," and a "tiny quivering hate machine." He ends by declaring, "a little dog doesn't want to be alive any more than we want it to live." But nowhere in the article did Ken explain the genesis of his hate-hate relationship with small dogs. I figured the piece had to be based in small part on experience and in large part on the writer's love of biting satire.

At least, that was my hope when I tracked down Ken to ask him about his decades-old scorched-earth excoriation of small dogs. A quick internet search revealed Ken now lives in the Mojave Desert and produces *Desert Oracle: The*

Voice of the Desert. The series includes a book, print magazine, and podcast show that's part desert podumentary and part darkly funny social commentary. I emailed him with the subject line "Small Dogs and Deserts," in the hopes he'd actually open an email from someone running a website called "The Small Dog Rules." Fortunately, my missive cracked him up, and we chatted via email.

Ken stopped writing for *Gawker* years ago, and he confided it always shocks him when his old articles surface. But given their wry look at American life, it's clear why they're still so popular. In what is clearly his trademark dry wit, Ken quipped he's "surprised that stuff is still online. I figure one day somebody will turn off the switch, to save electricity."

Well, they haven't turned the switch off yet, and since his small dog diatribe is still alive and beating online, I asked him (nicely) to explain himself. It turns out his small dog hate is rooted in his urban days living in San Francisco and traveling often to New York on business. It was the early years of purse puppies, popularized by celebrities I won't bother naming. Apparently toy breeds were invading Ken's space at every turn, and he says "suddenly these angry little dogs were screeching in every bar and restaurant and airplane flight."

He felt it was time someone called them on it.

And so he did, with his poison pen. Now, perhaps just to make me feel better, Ken concedes he "likes non-toy little dogs fairly well. Sometimes."

Whether his begrudging admission is genuine, Ken has a point: Small dog prejudice *is* based in reality. Just as there's no denying some breeds cause more injuries to

But Wait, What About Small Dog Syndrome?

humans, some small dogs are foaming-at-the-mouth, anxiety-ridden, get-it-away-from-me Chucky dolls.

Despite their beautiful ability to stabilize and improve *human* mental health, the entire small dog population bears an undeserved reputation for being imbalanced and neurotic themselves. It's undeserved because, like any stereotype, the exception often shapes the rule. In other words, the yappy, snappy behavior of our childhood neighbor's Chihuahua often inaccurately molds our perception of "small dog." And to the extent a small dog is actually neurotic, it's likely not because of failings in the dog but of the failings in the humans who've created these neuroses.

I speak from guilty, sad experience.

Later in this section (let's call it my confessional chapter), I go into painful detail about my fifteen-year sentence with Ruby, my puppy mill miniature dachshund, whose behavior nearly destroyed *my* sanity. Over her fifteen-year life, she flaunted every horrible habit people cite when they say they can't stand small dogs: incessant barking, house-breaking issues, nipping, and destructive and self-destructive behaviors.

So yes, I acknowledge some small dogs absolutely can be challenging, unlikable pools of emotional imbalance. (The same is, of course, true of some big dogs, some children, and the party guest you wish would go home. But that's another book.) However, now, I know everything "wrong" with Ruby resulted from gross human error—and nearly all that error was mine. Her tragedy began with her very birth to barn-housed dogs in a massive breeding operation. It ended with her human family's failures to find the right medical and training interventions to help her become the best dog she could be, despite her beginnings.

Ruby lived to be about fifteen and it's been seventeen years since she died. Stressful memories of coping with her many frustrating tendencies still haunt me. After her death, I swore I'd never have another small dog until my "old lady dog." That's the dog I'd always planned to adopt when I became too old to safely handle my beloved sporting breeds. But then, as I've shared with you, my goldens kept dying of cancer. We lost three in the span of about that same fifteen years, and I couldn't do it anymore. So, I began considering fast-forwarding the "old lady dog."

After spending several years studying breed groups, listening to dog trainers, and researching small dog science, I finally got it: Small dogs aren't the problem; their people are.

When we adopt small dogs from the right places (carefully vetted breeders or reputable foster/rescue organizations), and we avoid all the bad human habits that bring out neurotic Napoleonic behaviors in small breed dogs, we nurture all the loving companionship toy breeds can bring. When we purposefully engage with our small dogs by treating them not as "small" but as "dog," we will create one of the most enjoyable relationships of our lives.

Thus, while this book serves as a celebration of our relationship with small dogs, it is not a sales pitch designed to encourage everyone to own a dog. Indeed, as our shelter populations world-wide attest, a significant number of us aren't equipped—whether as the result of financial, emotional, or time deficits—to own any pet at all. Not even a fish. The following chapter uncovers the cautionary side of small dog ownership.

But Wait, What About Small Dog Syndrome?

THE VACCINE FOR SMALL DOG SYNDROME

Watch just a few episodes of pretty much any dog training show out there, and if you didn't know before, you'll know now: Small dogs don't know they're small. Or if they know, they don't care. While these tiny packages are undeniably cute, they can, if mishandled, also become anxious, domineering, territorial, or seemingly vicious.

In fact, in a 2022 study of dog breeds brought to a veterinary behavior center between 1997 and 2017, terrier breeds were "overrepresented" for aggression problems.[178] Another recent study showed that neither the Chihuahua nor the JRT were aggressive, but the research recorded higher aggression in smaller dogs overall. In this study, the small breeds most frequently mentioned were miniature poodles and miniature Schnauzers.[179]

Regardless of breed, however, these traits in *any* small dog are not simply unpleasant. They're dangerous to the health of the family's humans and to the well-being and care of the dog itself. The stereotypical yapping, nipping, evasive or neurotic behavior associated with tiny dogs is part of why so many people *think* they don't like them. Yet, ruling out small dogs based on misplaced bias is both unfair and foolish. While it may seem some small breeds may be prone to such personality defects, like *all* issues with *all* dogs, the problem almost always lies with the human owner.

Barring some genetic defect or illness affecting their brain or pain level, dogs will only do what we unwittingly bring out in them with our poor training. And with small dogs, the neophyte owner (and sometimes the weak or lazy or misguided experienced owner) will let them get away with figurative murder.

DOWNWARD SIZING DOG

Take Phoebe. Well don't, because you can't. We love her too much. And it's because we love her beyond reason that she, too, suffers from a mild case of Small Dog Syndrome, which we'll call "SDS" from now on. Truthfully, every negative SDS trait Phoebe possesses exists because we've let it. And the same was true of the occasional bad manners of our beloved golden retriever, Romeo. We know what we should do; we just sometimes don't do it.

Poorly behaved small dogs exist because the consequence of poor socialization in these petite pups generally runs in the vein of embarrassing but fairly harmless misdemeanors: irritating behavior, yapping, submissive peeing—as opposed to the large dog felonies of mass property destruction or life-threatening bites. Put simply, small-dog crimes are mere nuisances, for the most part. Thus, even long-time and allegedly wiser guardians can become lazy with our small dogs.

To reap the benefits of "dog" in a small package, you must treat your small companion not as a purse puppy, but as a full-fledged canine. You must create a culture in your home and life that makes your small dog part of the day's rhythms and that recognizes their essential existence as canines.

Just as we don't teach children in isolated moments and then let them run rampant at other times, every interaction with your small dog is a teachable moment. When the culture in your home is one of learning and positive reinforcement, then, and only then, will you have your cake and eat it too: You'll experience the joys of human's best friend in a package that fits seamlessly into almost any space.

But Wait, What About Small Dog Syndrome?

SMALL DOG CULTURE

Culture is important. It's important to humans, and it's important to animals. And it's especially important to dogs. They *think* culturally, sniffing out their family's benevolent leaders... and its weakest links. According to every modern dog behaviorist, dogs respond positively to consistent and loving leadership. In contrast, they react with fear, nervousness, and sometimes neurotic or aggressive behavior when surrounded by angry, punishing, unsure, inconsistent, or babyish treatment. Small dogs, especially, can become traumatized and develop behavior issues if treated harshly.

Thus, interact with your dog via canine, not human, culture. Although your small dog may be cuter (and smarter) than the neighbor's grandchild, treat your dog like a dog, not a baby. Yes, you can still cradle her like a baby, and talk nonsense to her, and smooch her on the nose when no one is looking (or even when they are), but most of the time, help her embrace being a *dog*, not a *human*. Know the key to a beautiful relationship with your small dog is there for the taking: Engage the "dog" and (pretty much) ignore the "small."

How do you ensure your dog remembers and enjoys her dogginess? Here's a non-exhaustive list, gleaned from my experience with my own small dogs and horror stories I've heard along the way.

THOU SHALT BAN THE PRIVILEGED BATHROOM

George Carlin had this funny bit that went something like this:

If aliens landed on the planet, they'd think dogs were in charge. Think about it. When you see someone following *someone else around, picking up their sh#$, the person picking up clearly isn't the leader.*

Every time I bag my dogs' poop, and I do it daily, I smile and shake my head. George was right. We are subservient to our pets.

I'm not advocating we abandon cleaning up after our pups, of course, but this story segues nicely into the First Small Dog Commandment: Thou Shalt Not Allow Fussy Elimination. That means:

1. Your small dog can walk.
2. Your small dog can walk outside.
3. Your small dog can walk outside in the rain.
4. And the snow.
5. And the sleet.
6. Your small dog can learn to potty outdoors.
7. Your small dog can potty outdoors in the rain.
8. And the snow.
9. And the sleet.

Most small dogs adore being outside when it's nice. Many little pups, however, turn tail (literally) and run back indoors at the sight of sprinkles or, heaven forbid, snow. Of course, a few toy or hairless breeds can't safely be outdoors in any extreme weather. But for most small dogs, hitting the brakes at the door when they see bad weather is simply silly.

Don't allow it. Dress your dog appropriately (if he needs a sweater, a coat, and boots with the fur to keep him warm, so be it), and then stare straight ahead, take a loose hand on the leash, and march right out into that gale as though it's just another beautiful day. If you don't, you'll be

But Wait, What About Small Dog Syndrome?

teaching your little dog that *he* decides when he potties outside and when he doesn't.

That's not a choice you want him making. Yes, it's hard when neither you nor he feels like braving the rain, sleet, or snow. And it's tempting to pick him up protectively when he high-steps around a puddle, shaking the damp off his little paws.

Don't.

No matter how cute you find such prima donna behavior, ignore it, and get yourself outdoors.

And bring the dog.

If your small dog senses your amusement, or worse, your stress, regarding her unwillingness to commune with messy Mother Nature, you're reinforcing her behavior. Instead, bundle both her and you against the elements, and make it fun. Run in the rain. Slide in the snow. But don't put up with a dramatic refusal to set paw outside. If you act as though being outside in any weather is simply what one *does*, she will do it. And then, you'll never deal with the SDS of bad weather bathroom accidents.

To be sure, many small dog owners swear by the small blue pads sold as puppy training mats, and when you're traveling in an airport or in a high-rise hotel, they can be invaluable. And yes, it's smart to train your dog to use them in a true weather or family emergency (we missed the boat on that). But generally, these pads should be used sparingly, and on your terms only, for several reasons, not the least of which is that your home shouldn't smell like a kennel.

More importantly, they shouldn't be an excuse for allowing yourself to leave your small dog without companionship for hours and hours. It's not fair. Plus, your house will smell like a kennel. But I said that. On to commandment number two.

House-training

Okay, okay. Let's talk about the elephant in the room. The one that pees where you don't want it to.

Small dogs have a reputation for being difficult to house-train.

In reality, most un-housebroken small dogs result from inexperienced owners who don't know what they don't know. They don't realize how miniscule their small puppy's bladder is, how vast their home seems to that puppy, or how much vigilance is required to see the signs their pup needs to go.

Small dog owners also feel less urgency when it comes to housebreaking. When a toy breed puppy piddles on the floor, you can dab up the tablespoon of liquid in moments. When a golden retriever puppy does so, you're scrambling to grab the paper towels before the pee river reaches your favorite wool rug. As a result, owners of small dogs often manufacture their own worst nightmare: an adult small dog who goes where she wants.

I know. I speak from experience. I've owned two small dogs in my life who were not reliably housebroken. Moses was someone's lost hunting beagle who burst into our lives when I was entering middle school. He'd likely spent most of his life in outdoor kennels, and he did not arrive housebroken.

He also never *became* housebroken.

He was my dog, and although I adored animals, I was eleven or twelve years old when we first adopted him, and I had no idea what I was doing in terms of house-training. Crate training wasn't a thing back then, and apparently no one advised us to try neutering him to prevent marking. At first, the problem wasn't major because when we lived on forty acres, he had the run of the farm and he nearly always wanted to be outside. When we moved to a small college town, however, and had only an unfenced yard, his days of baying at squirrels and running free were done.

That's when the real problem began. During the week, the entire family was gone at school and work, leaving Moses and Meg, our Irish setter, home alone all day. I would take Moses for a quick walk in the morning before I left for school, but I was a teenager and that "walk" comprised walking down the alleyway and back. My dad would come home at lunch to let them out, but Moses rarely concerned himself with waiting. So, scrubbing walls and carpet became a daily chore. I do not recommend it.

And then, there was Ruby, our puppy mill dachshund I adopted in my late twenties or early thirties. It was that period in our lives when we just couldn't house my beloved hunting dogs in our small city house and virtually nonexistent yard. That, and we had little time and even less money. I'll tell the full story of my abysmal judgment a bit later, but for now, just know Ruby's formative weeks spent in a crowded sawdust pen conditioned her to pee where she stood, even, or especially, when she stood on my carpet.

Of course, I can't blame the puppy mill entirely, as nearly every internet list of "Dogs That Are Notoriously Difficult to Housebreak" places dachshunds at the top of that list—either or both because their hound breed makes them highly distracted outdoors or their impressive scent abilities makes them return to the scene of their prior crimes, no matter how thoroughly you've scrubbed clean the evidence. Thus, Ruby's breed, combined with her birth factory's complete void of socialization and sanitation, and my lack of time and experience, dashed any genuine hope we had of making her a truly domesticated dog.

I bought training books. I tried every method.

We attempted crate training. But with shocking disregard for pain, she tore her mouth and bloodied her face in her mad quest to gnaw her way out between the metal bars. She'd work herself into a foaming frenzy in that open-air kennel, even when we sat in the room with her.

We attempted training her to use newspapers instead of our carpet. That was a short-lived experiment, as she took to pooping on the newspapers and then chewing them up, leaving a disgusting mess of soiled paper strewn across the sunroom floor.

Oh, and then, for good measure, she'd pee on the carpet.

I attempted taking her out so frequently she couldn't possibly have any liquid left to dribble in my house. She held it and dribbled anyway.

One particular night is etched vividly in my memory, though it happened well over two decades ago. Juggling a law practice and three over-scheduled kids who needed to be in three different places simultaneously, I always woke up before sunrise. In those days, Ruby had taken to peeing at night on the throw rug at the bottom of the stairs, which I would then throw in the wash every morning before leaving. It was at least manageable until I'd decide once again to try breaking the cycle.

In yet another desperate attempt to protect my home, I began taking her out before my ten p.m. bedtime and then I'd set my alarm to wake up around one a.m. to take her out again. I thought maybe she just couldn't hold her bladder, even though no medical explanation justified my theory. I figured I had nothing to lose but sleep.

But Wait, What About Small Dog Syndrome?

For a week, zombie-like, I'd sleep-walk from my warm cocoon to bring Ruby outdoors. After three days, it hadn't helped any—still the rug at the bottom of the stairs would be wet when we woke up. So, I'd throw the rug into the washing machine, sanitize it to within an inch of its life, and we'd replay the scene again the next day. After a week, my midnight forays with Ruby still weren't working, so I thought maybe it was the rug itself. I stopped putting it out, praying in the meantime that no child would take a header onto the unforgiving ceramic tile underneath.

It wasn't the rug.

She began peeing on the tile right at the bottom of the stairs.

I'd promised myself I'd give the nightly bathroom trips at least two weeks, thinking that seemed a reasonable time to help Ruby's body adjust to a new routine, where we limited her water after eight p.m. and took her out so much she'd have nothing left for the house.

On the tenth day, however, I surrendered. It was 1:15 a.m., and I'd just settled back under the covers after depositing Ruby back in the boys' room where she slept. I couldn't shut her in there, though, because her anxiety drove her to claw and scratch at any door in a closed room, whether or not she was alone.

Lying there awake only moments after returning from outdoors, I heard soft padding down the stairs. I stole out of bed to the upper landing and turned on the hall light. There, at the bottom of the stairs, was Ruby, peeing on the floor, only minutes after I'd taken her out.

And that was that.

Like a Cinderella without a fairy godmother, I picked up my bucket and pail and returned to my scrubbing.

But let's return to that elephant in the room—the reputation small dogs have for being difficult or impossible to housebreak. I just gave two prime examples. And in both cases, the common denominator was me and my gross human error.

With Moses, I forgive myself, as I was a child and lacked any true resources or ability to train him. And Ruby? Well, as Jessica of *You Did What with Your Weiner?* says, most dachshunds with house-training challenges are suffering from separation anxiety. I wish I had it to do over again, as I (1) would never have supported a puppy mill and (2) now know so many more medical interventions and positive behavior therapies that might have saved Ruby and me years of a sorry relationship. But in all cases, the fault was not hers. The blame for her lack of house manners lies squarely with the adult human who brought her into the world and with whom she lived.

So yes, it's *sometimes* true: Reliable housebreaking for some tiny dogs takes much longer than for large dogs. But that's not because these small dogs are less obedient, less clean, or less intelligent. On the rare occasions when small dogs take longer to house-train, it's because their bladders simply aren't yet big or strong enough to retain liquid for hours on end. Or they suffer from untreated anxiety. Those issues can be addressed if your eyes are open to what's really going on.

Thou Shalt Not Coddle Him

To coddle: treat in an indulgent or overprotective way.

Yes, small dogs are...well, small. So yes, they *do* require extra care against falls, predators, cars, rowdy children, household poisons, and clumsy adults. But, just as with children, we can protect them subtly while simultaneously helping them remain independent and tough. They deserve to live with confidence, and they can't do that if we're constantly hovering. Our own Phoebe is a perfect illustration of how easy it would be to ruin a perfectly magnificent animal, turning her into a neurotic princess. In fact, we came close to doing just that.

I'd watched *Animal Planet* training shows, read many training books, and worked with animal trainers throughout my life.

I knew—I know—all the "shalt nots" associated with the anti-coddling commandment: Thou shalt not consider your small dog to be a baby or a toddler. Thou shalt not carry your puppy without good reason. Thou shalt not encourage your small dog to fear things.

You get the idea.

And all these shalt nots were remarkably easy to enforce with all my big dogs…especially my golden retrievers, who positively lived and breathed to please. They not only learned to go to the bathroom outside at a mere ten to twelve weeks old, but I swear they would rather have died than go to the bathroom indoors. Likewise, they didn't expect me to carry them down the stairs or over puddles because they were big enough to handle doing so themselves.

But, because picking up a ten-pound fluff ball and walking her downstairs or having that sweet fluff ball's face peer innocently at you as she stands on two legs, leaning against you, is really no biggie, we often don't teach our small dogs basic manners. And when I say "we," I am including "me."

My mistake with Phebes occurred the horrific moment I watched her belly slide uncontrollably down a flight of stairs. Just days before, I'd taught her to navigate our hardwood stairs to the second floor of the Colonial we lived in at the time. The steps were steep, and there were thirteen of them.

The unlucky number should have served as a warning to me.

I knew from watching and reading training advice that I shouldn't be carrying my small dog around everywhere. I should expect her to move from point A to point B of her own volition. And so I did. But she was only ten weeks old and neither she nor I understood maybe she wasn't ready to be a big girl on the stairs.

The morning she somersaulted to the bottom, she was following her big brother Romeo down the steps. Since he always did everything at breakneck speed, and she, in typical little-sister format, copied everything he did, she also took the stairs at similar speed. I watched from the top as her bottom legs overtook her front legs and she lost her footing. Spread-eagled, she slid, as I let out a strangled "no, no, no, no, no, no, no" for every stair her little fat belly hit on the way to the landing. She hit the bottom, not terribly hard, and stood up, with me in hot pursuit to scoop her up and check for certain dire injuries.

There were none.

So, what I should have done was to help her trot back up the stairs and back down again. What I did instead was to freak out, and then carry her around all day reassuring myself every half hour that she could, indeed, still bend her limbs.

And from that day forward, every time we approached those particular stairs, someone had to carry her up and carry her down. Never mind that she approached any other stairway with utter fearlessness. She had learned the center hall staircase was an Everest to avoid.

But Wait, What About Small Dog Syndrome?

She had learned that because I taught her.

For a full year after that, we were her human escalators. And then one day, after she'd happily climbed a particularly steep set of stairs at my sister-in-law's house, and after I'd dropped and shattered a favorite coffee mug trying to carry her up the dreaded stairs along with an armful of other miscellany, I decided she could darn well learn to tackle Everest. Of course, I also knew I was a large part of the problem with Phoebe's fear of these stairs; after all, I'd started the whole debacle when I'd panicked and then began ferrying her up and down myself.

Rather than trying to undo my own damage, I enlisted the help of Leah Brocato, who was in her early teens at the time. Leah is the daughter of a good friend and colleague, and I'd long admired her deep regard for animals. Her menagerie includes three dogs, a horse, and a rooster named John Snow along with his hen harem: Daenerys, Sansa, Ygritte, and Cersei. For years, I'd watched Leah work with her animals, using self-taught positive training strategies. In her deep connection to four-legged beings, she reminded me of myself at her age, though she was much more accomplished an animal handler than I'd ever been. So, it was to Leah I turned when I got serious about ensuring Phoebe's confidence on those center hall stairs.

I thought it would take weeks. I thought wrong.

Leah simply brought Phebes to the bottom of the stairs, some of her favorite treats in hand, and lured her up the first several steps. She then lured her down. They then added a few more stairs, and then a few more, and within almost no time, Phoebe trotted right ahead of us, bebopping up and down those stairs like she'd done it alone every day of her life. In the space of less than fifteen minutes, Leah had cured an issue I'd spent a year creating.

Since Leah's fifteen-minute fix, Phoebe races up and down all stairs without hesitation. She is a regular dog, not a toy to be carted around and coddled. So don't make the mistake I did, which is natural but not good for your small dog. Unless there's genuine danger (or you're traveling and trying to avoid mud), do not pick her up and carry her places. Let her move from point A to point B with dignity and under her own power.

THOU SHALT NOT ALLOW HER TO BE A BULLY

Our sweet, ten-pound Havanese Scout exudes "cute." Her soulful, chestnut-colored eyes, her sweet puppy ways, and her silky hair in every shade of copper combine to touch some ancient protective instinct in anyone who meets her. When she stands on her hind legs and lifts her front paws and legs toward you, in the universal human toddler language that means "pick me up, I love you," she pretty much owns you.

In short, every cell of Scout's being radiates the adorableness that activates fierce parental feelings—those mama lion urges driving me to throw myself in front of anything or anyone who might hurt her. Scientists, always looking for more poetic ways of putting things, dub that a "neotenous" connection. Apparently, we're more attracted to and exhibit greater care over creatures with baby-like or "cute" features,[180] and we're likely to see our protective reactions thrum when a small animal stands on two legs.[181] And with Scout, we suppose it's true. She is so darn cute she makes our hearts hurt.

We strongly suspect she's privy to this science, given her habitual way of gazing into the eyes of any nearby human, jump-starting that oxytocin loop we talked about in Chapter Four. And that's one reason small dogs are so likely

But Wait, What About Small Dog Syndrome?

to be ill-behaved. They capitalize on cute, and the chemical reaction they create for us with that oxytocin loop raises a human's happiness. It's nearly impossible to fight.

Phoebe was the same way as a puppy. She easily inspired melting feelings of love and adoration.

But she was born with the heart of a fearless viper.

At ten weeks old, she was so full of her own importance we quickly christened her Phoebe Badger, inspired by the viral "Honey Badger" YouTube video.[182] She was a menace to Romeo, to anything that moved outside, and to herself. But, darn, she was so adorable doing it.

It's at this point many small dog owners veer onto a path it's hard to double back on. It's the path where you ignore the emotional (or actual) bullying at which some small dogs are so adept. Just as we teach toddlers that manipulation isn't fair play and hitting to win your own way is unacceptable, we must do the same with small dogs. We must quash the urge to enjoy the adorableness of these sometimes-fierce little beings or eventually we'll create a true bully.

With Scout, her specialty was, and still is, emotional manipulation. She would have you rubbing her belly twenty-four/seven if she could. Or feeding her tidbits from the table. Or paying attention to her instead of working. While, at times, it's good she reminds us of what is important in life (by smacking us incessantly with her paw), at others it's not. So, we made it clear "leave it" and "go lay down" are concepts she must embrace.

With Phoebe, her behavior was mostly directed against Romeo. She'd hang on his tail or his magnificently feathered chest...anything to get him to play. Or she'd attack autumn leaves or the neighbor's trash can that suddenly loomed in their driveway on trash day. It was all very cute, except it wasn't. Not really. If we'd let her bullying behavior go on, she'd have tortured Romeo. If we'd permitted her to lunge at every stray item in her path, she'd eventually have hurt herself or someone else. Thus, we made it clear "be nice" means to stop your grappling with whatever's in front of you.

From the beginning, small dogs need to be taught they don't own the world and others deserve the right to be left in peace.

Thou Shalt Expect and Reward Good Manners

Closely related to our anti-bullying stance is the concept of good manners, and this is where so many small dog owners lead their pups astray. As we've said in the past, and as science has confirmed for us, small dogs have a reputation for bad behavior. And yet, the researchers who uncover these findings also hypothesize such misconduct might be laid at the feet of humans who tolerate or even encourage it.[183] In short, families with small dogs often fail to train them, because they mistakenly believe their inappropriate behavior isn't that big a deal.

Until it is.

We've already discussed house-training and politeness. The other major complaint against small dogs is their tendency to yappiness. And some breeds are more prone to it than others. Our two Havanese would bark incessantly if we hadn't taught them the concept of time and place. And then kept at it. Just as you can't rely on one lesson to ensure your toddler doesn't run into a street, you can't count on your pups to listen reliably after being told only once that barking their heads off isn't a thing in your home.

No matter how quickly a dog or a child learns what you expect of them, they're not always going to deliver. Phoebe and Scout learned the *word* "quiet" within one afternoon. But learning what a word means and actually applying that word to themselves, consistently, when they're not in the mood to listen, requires an ongoing conversation with them. We must continually help them connect that "quiet" vocabulary word with our family's concept of time and place.

    I reinforce the lesson on the daily, as we shared in previous chapters. They do their due diligence, alerting us to potential intruders on the horizon. We thank them, say "quiet," and after some under-the-breath backtalk (usually from Scout), they settle back down. Occasionally, I need to stop whatever I'm doing, walk to the front door or slog up to their third-floor perch to look them in the eyes and politely ask again: "Quiet." And then I stand a moment, leaving a brief hush in the air before praising and thanking them. Do I enjoy having to interrupt my focus and rise to chat with them face to face? Well, no. But I've learned to appreciate it, embracing it as an opportunity to move. And I know it's vital to their continuing education.

    Such gentle, nonstop learning means their understanding is so complete now the girls listen no matter who's talking. Whether it's a young neighbor, a dog walker, or a visiting friend, Phoebe and Scout hear what their surrounding humans are asking. We once had an elderly relative staying with us who liked to take an afternoon nap. On one day during her visit, I used her rest time as an opportunity to run overdue errands, but I worried the girls might bother her by barking while I was away. So, I shared with her the "okay, thank you, quiet" routine. When I returned home, our friend related with such pride how the pups "hushed right away." And she swears they understood she was napping.

Of course, that level of human/canine communication takes several years of total consistency. Even now, if I skipped a day of thanking them and letting them know they'd watchdogged beautifully, they'd likely revert to barking their heads off. But the payoff for that consistency? Immeasurable.

And the pain if you fail to train and reinforce that training is just as immeasurable. I witnessed such failure just recently, after a family had set themselves up on the little beach near our house. Initially, their young mini doodle was playing on the beach with them, but then he became too enamored of digging in the sand, and they moved him up to the grass, secured by a long lead.

Alone. While everyone else in the family was enjoying digging, splashing, laughing, and having a wonderful time. His response? Sharp, incessant barking. Phoebe and Scout ran out to our deck to watch, pacing a bit and sharing this fellow creature's obvious anxiety. His piercing barks continued at least twenty minutes, echoing up and down the shore, disturbing everyone's peace and likely making everyone who didn't know better shake their heads and mutter something about "yappy little dogs."

Yet we've learned it's not the dog, is it? He was set up to fail: isolated but within sight of his family, removed from the fun, and given no distracting thing to occupy him. To top it off, he received no direction insisting his barking be replaced with some more acceptable behavior, like enjoying a frozen Kong or a puzzle toy, or heaven forbid, playing with his humans. Someone finally resolved the matter by unleashing him, picking him up, and carrying him into their friend's house. There, his small dog voice was muffled by the walls, no longer disturbing his family outside.

But Wait, What About Small Dog Syndrome?

But that didn't really resolve matters, did it? The dog didn't learn a thing about when barking is okay and when it's not. So, the next time he's out with them and not getting his way, he'll bark. And they'll put him away. And they'll keep doing that until eventually, maybe they won't bother trying to bring him in the first place.

We see similar issues with small dogs who aren't consistently taught "sit," or "leave it." Because toy breeds are so easy to move and cause little damage, relatively speaking, misguided owners don't provide them with the tools they need to operate in the larger world. Their fears or foibles are ignored, or worse, encouraged, and as adults, their traits can become magnified and unmanageable. But if we teach small dogs with the same high standards we set for large ones, we build the confidence and well-mannered behavior that brings out the best in them.

THOU SHALT INTEGRATE YOUR SMALL DOG INTO YOUR LIFE

Though I talk repeatedly throughout these pages about how seamlessly small dogs fit our lives, many small dog owners don't do what's necessary to help that happen. Some of us just don't play with our small dogs like big dog owners do, nor do we obedience train them. At least one group of researchers backs up my theory that we treat small dogs differently than we would big dogs because we can—we don't fear their bad behavior.[184] And yet we should fear it, not merely for our own sakes, but for the sake of the dignified little beings who didn't ask to come live with us.

Small dog socialization is a lifelong process. It's vital for your dog's mental health and your family dynamics. And while it's infinitely easier just to hop in the car and leave them behind, don't. You won't regret it. On one recent visit to Home Depot, Ron and I brought the dogs, as we often do. It was a quick trip—we were in and out in fifteen minutes. Was it worth the extra ten minutes we spent harnessing the dogs and taking them in and out of the car, and the bit of additional care we took in the store, guiding them away from potential dangers or non-dog-lovers?

Yes.

Why?

So many reasons: the smile they put on the faces of at least six toddlers strapped into shopping carts; the bonding we created for a mom and her two teen daughters who stopped bickering over how much longer they'd be there long enough to bend down together and pet the girls; and the friendly exchange with the random-stranger-in-a-hurry who nevertheless slowed enough to admire them and say, "great dogs." They left nearly a dozen smiles in their wake.

And that list compiles what Phoebe and Scout did to spread happiness to the world in general in just a single outing. What about the shared joy and daily memories they provide to Ron and me? Rather than a mundane hardware store run, we enjoyed the little rituals we've built as part of our family: the joking reference to someone mislabeling Scout "the brown dog," the smiles we exchange as we both enjoy the way Scout proudly leads the pack but obsessively looks back to see we're still there, the way the window needs to be cracked just so in order to ease Scout's carsickness, and our pride and gratitude for Phoebe's fondness for and gentleness with children.

We're so proud, in fact, you'd think we birthed her.

But in a way, we did. Because we include the girls in what we do much of the time, they know how to be in the world. They are powerful beings in small dog packages. And that has made all the difference.

CHAPTER TWELVE

A Brief Bit of Advice About Finding Your Small Dog

Downward Sizing Dog is a defense and celebration of small dogs and their beautiful canine ways. My next book (tentatively titled *The Small Dog Rules*) will delve more deeply into the "hows" of parenting small dogs, perhaps including a detailed look at the process of finding a small breed puppy or adult dog. But here, it's well beyond our scope to cover those topics. Having said that, because so much of one's happy relationship with any dog hinges on how everyone meshes into family life, I added this chapter to provide a brief bit of advice regarding puppy mills and breed groups, in the hope you might learn something from my mistakes.

Many of you will think you won't need this chapter, as you're going to find your small dog through your local shelter or reputable rescue. All three of my adult kids created their dog family this way: Drake (my deaf and daring Jack Russell/bull terrier mix granddog), Brady (my protective little granddog of very uncertain ancestry), and Rosie (my Chiweenie granddog who learned to walk again after her eight-thousand-dollar emergency spinal surgery) round out our family beautifully.

If you, like my kids, plan to rescue a dog, please stick around for the subsection here on breed groups. It contains information that might help you decide which shelter dog would be happiest in your home. Then, read this chapter's last subsection: "The One Thing." Recent research shows that up to twenty percent of rescued dogs are returned to shelters within the first six months of adoption.[185] Obviously, such returns are heartbreaking for families and truly tragic for the dog, who now may appear less adoptable to others. You want to eliminate the chances of that happening. Thus, "The One Thing" subsection might help you focus on what's most important when adopting your pup.

If your heart is set on a puppy, then you'll want to read what I have to say next, because puppy mills prey on small breeds. It's not a pretty story.

Avoiding Small Breed Puppy Mills

When searching for a small dog, you're going to run across unethical breeders. While large breed dogs can also come from puppy mills, someone breeding for maximum profit is likely going to prefer a "product" that's easier to deal with. Thus, someone breeding for money is going to market to small dog fans every time, and here's why: Small breed dogs can be crammed into tiny spaces, it costs comparatively little to feed them, they're easier to clean up after, they keep their cute factor even as they become older, and they're cheaper to ship.

That all translates to a much higher profit.

A Brief Bit of Advice About Finding Your Small Dog

And it's so very easy to jump into their trap. Puppy mills specialize in slimy (and often fraudulent) but highly effective marketing promising you the exact rare small dog, in the precise gender and color you want, no waiting required. Here's where my continued story about Ruby might help the concept come alive for you.

When I was in my late twenties and early thirties, I thought I knew everything about dogs.

It turns out I knew nothing.

Or at least, very little. In particular, I knew nothing about puppy mills, which are essentially puppy factories. They're horrific places that breed for money and not the health or well-being of the pups at their mercy. They treat their dogs like livestock or worse. They churn out puppies born to exploited female dogs kept in inhumane and abusive conditions. They offer one-stop shopping, so they're marketing genius, appealing to our love of convenience and speed. And I wanted both during a particular spring back in 1993, when pregnancy hormones had temporarily derailed my common sense.

I was in my very early thirties. I had two full-of-life boys under six years of age, I was weeks into an unexpected pregnancy with my little girl, and I'd recently opened a solo practice law firm. Somehow, adding a puppy to the mix made sense to me. Like I said, hormones and the shock of unplanned pregnancy stole my judgment. That, and I am a human barely capable of living without canine companionship.

Against the protest of my (saner) spouse, I scoured the newspaper ads (this was before the days of endless internet sites packed with puppy faces) and found a kennel in a neighboring state selling many dog breeds. They had dachshunds, which is the breed I had chosen, believing their combination of "small" and "hunting dog" would be a perfect compromise to satisfy the reality of our lives. That reality included a virtually non-existent yard and nine- or ten-hour workdays. So, one Saturday morning, I bundled my three- and five-year-old sons in the car and headed for the rolling hills of Massachusetts, envisioning an idyllic farm where carefree puppies rolled happily in the grass.

What I found was not that.

The owners directed me to a large barn where dozens of different puppy breeds and their moms were held in separate pens. Since this portion of the puppy factory was open to the public, I'm not surprised to recall it being relatively clean, with only a faint zoo smell wafting up from the sawdust floor. We were asked no questions about our experience with dogs or what type of home we'd provide, and they were allowing puppies to leave their moms at a mere six weeks of age. They took my check, and we were soon on our way home.

That was over twenty-five years ago.

And yet not much has changed, except the mills are more sophisticated now. Setting aside the inherent cruelty of such operations, it's simply inviting chaos if you bring home a puppy from a puppy mill. Dogs from commercial breeding operations overwhelmingly experience terrible beginnings, and a recent international survey of dog breeders noted puppy mill and pet store dogs show more behavior issues (including fear and aggression) and are surrendered more frequently.[186] As we discovered with

A Brief Bit of Advice About Finding Your Small Dog

Ruby, they also suffer from significant anxiety and housebreaking issues.

When you're sure you want a small dog, whether you go through an adoption agency or a breeder, a rescue, or a shelter, it's all about finding a dog to fit into your life. It's also about avoiding anything like the fifteen years Ruby and I spent together. That means you must arm yourself against savvy puppy mill breeders whose facilities appear pristine but whose underbelly, if revealed, would show dogs with poor genetics and a substandard, if not downright abusive, quality of life. To help you avoid inadvertently supporting such awful places, I've created a downloadable checklist of puppy mill warning signs, and you can access it at www.smalldogrules.com.

Understanding Breed Groups, and the Smallest Dogs in Each

If you decide to expand your family to include a small dog, the key recommendation I can make to you is this: Understand what your daily life truly is and what you want it to be. Consider family personalities, lifestyles, and know the basic personality of the pup you'll be making part of your life. And, understand your pup's breed group or groups and their general way of interacting with life. Obviously, that's easier with a purebred dog and more challenging with mixed breeds.

Fortunately, most reputable shelters and rescue groups can help you identify the likely parentage of any mixed breed pup you want to adopt. More importantly, good breeders and dog adoption organizations can also discuss whether your potential pup's personality fits the mold of his breed groups. In fact, you know you're in the right place if you're being educated about all the above, as I experienced when I went to pick up Phoebe.

The moment we met, I knew this puppy was a force to be reckoned with. I'd driven thirteen hours straight from Rhode Island to Ohio, and I was exhausted but eager to meet both her and her wonderful breeder, Karen Warncke.

Before I was in the door, little five-pound, five-inch-high Phoebe pranced over and gazed up expectantly, as though she believed my role in life was to entertain her. She moved immediately into a real downward dog, splaying her front paws, and wriggling her hind end in the air. She accompanied that pose by looking me straight in the eye and sounding a sharp, insistent bark.

No shyness or submission here.

I clapped my hands on the carpet next to her, and she sprang into action, tackling my open palm.

It's at moments like these they hook you. It's almost cosmically ironic, in a wonderful, heart-breaking sort of way, that such a small mite of a thing can take herself so seriously. She and I played, Karen and I chatted, and I was smitten.

We spent about an hour together, while Karen shared stories of both Phoebe's parents and her puppy antics. I'd seen pictures of the entire litter, and though admittedly I was biased, Phoebe was definitely the cutest among a litter of adorableness.

But I had a question: Why had the other puppies been adopted before her?

Without hesitation, Karen volunteered, "Well, she's kind of ornery."

That little fluff ball in front of me? Ornery? My face reflected my question so clearly, I didn't need to speak, and

A Brief Bit of Advice About Finding Your Small Dog

Karen quickly explained this last-of-the-litter puppy was not your usual Havanese.

You see, Phoebe was fiercely independent and less cuddly than most Havanese tend to be.

For us, that was perfectly fine. We had our golden boy Romeo, who showered us daily with so much "love me" vibe that our daughter Caroline used to quip "he just doesn't know what to do with all his feelings." Thus, we could welcome a pup who valued her alone time. I assured Karen that Phoebe's independent personality would mesh beautifully with our family. Many breeders would leave it at that, take my check, and hope not to see me again. But that is not the way of a principled breeder, and that is certainly not Karen's way.

Instead, she touched my arm as I walked out the door. Looking me in the eye, she said, "If you get her home, and you discover she's not what you need or want, you just bring her right back. We'll figure it out. I want you and her to be happy."

As our story with Phoebe demonstrates, not all purebred dogs fit their breed group's intended personality. The same is true of mixed-parentage pups. Indeed, you'll recall there's emerging research showing dogs are individuals, and it's unwise to make behavior assumptions based on a pup's breed.

And yet, it is still crucial you understand the genetics that might, at least in part, drive the personality of your potential new family member. Here, I'm inviting you to dip your toe into the vast information available regarding the psychology and genetics of small dogs. That psychology may well differ depending on the *breed group* to which your potential family member belongs.

Not the breed. The breed *groups*.

Because it's the umbrella breed group that often matters most. No matter where you get your dog, and regardless of whether your pup is a purebred or a mix, knowing terriers can possess relentless prey drive or small hounds are prone to taking off when trailing a scent is vital information for potential pet parents. Don't let anyone sway you from gathering this knowledge. Too many dog trainers and so-called experts encourage potential new dog parents to ignore breed information. But in doing so, they're asking you to ignore science and hundreds of years of humans purposefully aiming for certain traits in breed groups.

Don't do that.

Multiple recent studies looking at the genetics of breeds confirm what seasoned breed enthusiasts have long known: At least some of a dog's behaviors—including how easy they are to train, how likely they are to chase, how assertive they are to outsiders, and how much of the spotlight they seek— have a distinctly genetic base seemingly connected to breed.[187]

Having said that, at least one recent groundbreaking study found AKC breed stereotypes really don't play out, and trainability is the characteristic most attached to breed.[188] But remember what we're talking about here are the major breed *groups*, not the breeds themselves. By exploring each group and understanding the many small dogs in these varying groups, you'll gain clarity regarding your own small dog's psychology and some of the DNA that drives it.

Most significant for the everyday dog owner is the fact that the very traits that fit your life and make a dog either easy to live with, or not, are the traits in which genes call the shots sixty to seventy percent of the time. At least according to current science.

The internet is packed with great resources to help you develop a thorough understanding of each breed and the group to which it belongs, and we'll be including a list of our favorite resources on our website. For our purposes here, I'm giving you an overview and a shove in the right direction to discover more.

So, what, exactly, are breed groups? Put dryly and technically, breed groups are an agreed upon categorization of dogs that brings varying breeds together under one umbrella according to a dog's original purpose and function. When dog shows first gained popularity in the 1800s, the original groupings were simple: sporting and non-sporting. Then, the terriers and toy groups spun off, and so on, until we ended with seven major groupings, which we talk more about next.

Why are breed groups important?

Because you want to know who you're going to be living with.

Dogs in most breed groups have been bred for centuries upon centuries to carry particular drives. And those drives are called "drives" for a reason—they quite literally can drive your dog to do things you wish he wouldn't do. Expert trainers aside, a dog's breed group makes up a large part of a dog's personality and people skills. So, it's nice to develop some background in them.

Globally, seven major kennel clubs dictate the rules, and sometimes disagree, regarding the requirements for a particular breed and which breed belongs in what group. For our purposes, it's not terribly important to understand the rules of competing kennel clubs, but once you narrow down the breed group you like, it would be worthwhile for you to explore how other countries' breed clubs characterize the group.

In the meantime, we'll work with the AKC, which has categorized seven breed groups: terrier, toy, working, sporting, hound, non-sporting, and herding. Of the seven breed groups, three of them (herding, sporting, and working) list *no* members under twenty pounds—the cutoff we've selected for a dog qualifying as "small." That leaves us to study the four remaining groups: terriers, toys, non-sporting, and hounds.

THE TERRIER GROUP

The word "terrier" derives from Latin "terra," or "earth." Dogs in these groups were literal "earth dogs," driven to root out nasty vermin like rats and mice. Modern Americans now use friendly pest control companies to do such work, so we don't depend on (nor do we probably want) our dogs to bring us any bloodied little mammal bodies (no offense, Daisy).

If you're thinking of a terrier, read everything you can about the group, as they aren't dogs for the laid-back or faint of heart. As nearly any terrier website will tell you, many of these little dogs are driven and, when in their focus mode, they're borderline belligerent. They are full of spunk and vinegar, and they can be so much fun.

But they need a compatible owner.

Terriers include dogs of widely ranging size, from the smallest Norfolk terriers — weighing in at eleven to twelve pounds, to the largest—the Airedale terrier, with males weighing in at between fifty and seventy pounds. Of the thirty-plus terrier breeds in the group, twelve of them are under twenty pounds. I'll be putting together a list of those twelve, which you'll be able to download at www.smalldogrules.com.

THE HOUND GROUP

Dogs in the hound group are trackers who use their superior sense of smell or sight to hunt other animals. They were bred to work on their own, for the most part, and can be very driven by that instinct to chase down other animals. Of the over thirty breeds in the hound group, only three weigh in under twenty pounds: the beagle, the dachshund, and the Portuguese Podendo pequeno. I've been owned by two of the three (the beagle and the dachshund), and I can attest to the independent nature of both these pups.

Despite their self-reliance, both dog breeds are deeply affectionate. At least ours were. Beagles are incredibly sweet and undemanding…unless you stand between them and that rabbit, or squirrel, or anything else that moves. And dachshunds? Well, you know most of my story with dachshunds. But again, that was my fault, not Ruby's. And my boys worshiped and adored her, my oldest crediting her with having saved his sanity in middle and high school. Having said that, unless I had no other option at all—unless a dachshund was the last dog on the planet—I'd never have another one (please don't tell Jessica). I am in the distinct minority there, though. And I'm surely misjudging Ruby based on my bad choices and abysmal training.

And yes, I'm collecting pictures and personality traits of these three dogs which you'll be able to download at www.smalldogrules.com.

THE TOY GROUP

All the pups belonging to the toy group are under twenty pounds. That's why they're called toys, although, frankly, someone needs to tell the global kennel clubs they're doing a marked disservice to most small dogs by continuing with this outdated view of what a small dog is.

They're dogs, not toys.

Since most breeds belonging to this group were meant for the primary purpose of providing human companionship, I'd suggest we call them the "companion group." As a whole, the many dogs in this group are highly engaged with, and attached to, their humans. But here, I must give at least some credence to those trainers who say that breed groups don't matter. I will agree that in rare instances, a retriever won't retrieve, a hound won't hunt, and a toy breed won't be terribly interested in giving you rapt attention.

We experience it in our own Phoebe—the pup whose breeder suggested she was "ornery." Years later, we enjoy daily proof of Karen Warncke's knowledge not merely of her Havanese breed but of each puppy in her line. Indeed, she knows her individual dogs almost as well as she knows herself, we think.

Our grouchy Phebes is now seven years old, showers us with affection only between the hours of eight and ten a.m., and then prefers playing, watchdogging, and neighborhood walking patrol to any form of cuddling. She leaves that to her red-headed little sister, Scout. Scout is also from Karen's Havanese Treasures lines, and she is a Havanese poster child: funny, mischievous, and madly affectionate always.

So yes, it's *possible* you could wind up with a dog who follows few or none of the behavioral characteristics our

A Brief Bit of Advice About Finding Your Small Dog

forehumans purposely bred into them. Yet on balance, your pup is likely to carry at least some traits of its group. So why not work *with* nature when you can and not against it?

If you're interested in a pup whose primary purpose is to be a friend to humans, visit www.smalldogrules.com to download the twenty-plus pups in the companion group. Yes, I'm trying to make it a thing.

THE NON-SPORTING GROUP

Who names these groups? "Non-sporting." Really? This tells us nothing. And speaking of nothing, that's what the dogs in the non-sporting group share in common: (almost) absolutely nothing. Their only connection is, like all the other dogs in this group, they don't fit anywhere else.

As a result, the non-sporting group is the one place where you won't learn anything important in terms of personality traits. Instead, I recommend reading the history of the dog breed you're interested in and determining the dog's origins. Many times, that allows you to understand whether the dog is likely to have high prey drive, like most terriers, or an unstoppable instinct to follow a scent, like many hounds. Of the twenty-one dogs in the non-sporting group, nine are under twenty pounds. And yes, you'll be able to download the list on our website.

THE ONE THING

The ever opinionated "they"[ii] say there are no guarantees, nothing is ever clear cut, there's no right or wrong, and things aren't set in stone. Nowhere are these cliched reminders truer than when we're exploring the

---

[ii] Who are these people, anyway? And why do they hold so much sway?

personalities and behaviors of the beings who share our lives. Just as with our human relationships, it's never just about the *other* being. No—it's very much about *us* and about an honest inventory not merely of our own needs and wants but of our strengths and limitations.

But before we dive into angsty self-reflection, let's first look at the other half of the equation in any relationship: the half that's not you. Even if you're a long-time dog owner, before you rescue a dog or bring home a purebred puppy, re-educate yourself. Become a self-taught expert on *all* dog breed groups and even dog breeds. Read. Contact breed associations and speak with experts. Chat, and then chat again, with the rescue people who know most about the dog you're considering adopting.

Because, while yes, the science is clear that canine breeds don't *necessarily* behave the way they're "supposed" to, it's also true that hounds lead with their noses, sight hounds pack speed, and terriers, well, mice and all other small mammals beware. If you understand these possible personality traits before entering a lifetime relationship with your dog, you'll make wiser choices.

And now let's look at what's arguably the more impactful part of the formula: ourselves. While I've shown throughout *Downward Sizing Dog* that small dogs are powerful beings capable of making our lives better, they depend on us to do the same for them. Remember, small dogs are like human toddlers in their consciousness. They require generous and loving human care, interaction, and socialization. We can enjoy the transformational perks of small dogs in our lives only if we're first able and willing to nurture and guide them—to make them true family members.

A Brief Bit of Advice About Finding Your Small Dog

Thus, part two of the equation requires us to evaluate honestly whether we're equipped to do that. If we are, well then, welcome to a world where the small dog rules.

EPILOGUE

THE SMALL DOG RULES

My purpose in these past twelve chapters has not been to prove big dogs are bad or small dogs are the best choice for everyone. Well, maybe the second part, just a bit.

Downward Sizing Dog isn't about size shaming. Indeed, of any good facets I might possess, my very best sides were shaped by the vibrant big dogs who enriched my childhood and most of my adult life. But I confused their large size with their canine worth. I somehow assumed it was my dogs' physical heft that marked their inclusion among my truest friends. Looking back, however, I see my mistake: It wasn't their bigness that made them important dogs or that generated their impact on my life.

It was their essential nature as "dog."

That vital canine spirit abides in my small dogs in the same way it existed in my Irish setter and my goldens and every large dog who's impacted my life. It was and is their boundless glee over the basics—food, water, walks, and warmth—that taught me to keep my eye on what truly matters in the moment. It was and is their steadfast, wordless emotional support that proves talking doesn't cure everything, and sometimes silence and journeying within is the strongest path. It was and is their companionship in our daily outdoor quests that fuels my awe of and deep attention to nature's transformative forces.

In short, all my dogs' joyous interactions with every day nurtured my purposeful optimism, my grounding in the present moment, and my healing bond with nature. These three parts of me, formed through my constant companionship with dogs, are my bedrock. And this foundation ensures I don't merely survive the daily life stressors we all face––I carry a core strength that enriches my days and permits me to lend that strength to others.

Thus, downward sizing dog isn't about elevating small dogs above their larger siblings. In fact, I'm asking we do the opposite—that we look beyond the physical and recognize the essential soul of every dog, no matter the size. The truth is each and every dog can make us better humans, if we're willing to listen and learn.

But do remember: Small dogs are dogs.

The End

DOWNWARD SIZING DOG

Scout stands on two legs, her eyes wild, her teeth bared in wolf-like imitation as she leaps on Phoebe's head. Phoebe catches my eye and rolls on her back, allowing her smaller sister to overpower her, letting Scout believe she's actually the one in charge. They growl and wrestle for a good five minutes but keep looking to me in the midst of their tussle, wondering whether I'm going to close the stupid laptop. I'm sure that's what they call it, in their heads. It's their nemesis, I think.

And I? I am in shock. For today, we are done. Actually done *with this book that took five years and occasionally wrote itself.*

And often didn't.

I say "we," because as Ron can tell you, Phoebe and Scout worked with me every keyboard stroke of the way. First it was Phoebe sitting next to me at five a.m. as I tried to eke out five hundred words before school. And then Scout joined us, and they were my shadows, changing positions from the sofa, to their bed under my desk, to the kitchen island, following me around wherever I worked.

People say writing is lonely work, but not if you have a dog.

A small one.

Or two.

It's a beastly hot August day and the weather guy says it's only ninety-two degrees but feels like ninety-nine. No matter. The girls are always up for paddleboarding, the two sometimes sharing the tip of the board where they enjoy the best view and sometimes knocking the other in the water when apparently only a solo lookout will do. The swim will do us all some good. Plus, they deserve the celebration.

If it weren't for them, life would be so much less than what it is.

Acknowledgments

It's impossible to articulate the impact my husband Ron had on *Downward Sizing Dog*'s creation, but I'll try. Ronnie, you are my truest friend and wisest guide. You supported me whole-heartedly while challenging me at every turn, and we managed a book in the end.

My children were instrumental in this work: Jackson, your creative brilliance, given so generously when you had little time to give, was key to helping me shape *Downward Sizing Dog*'s message; Nate, your mom check-ins never failed, and your logical, cut-through-the-drama advice was always spot-on; Caroline, despite your own demanding career as a hospitalist in the midst of the pandemic, you always provided gentle encouragement and an unquestioning belief that this book would be great.

Thanks to my extended family, who for five long years genuinely believed in me and listened to my answer when they asked, "how's the book coming?" That sincere interest meant so much.

Deep gratitude goes to my editor and colleague Jen Steffy. You took an unwieldy manuscript and helped me reconfigure it into an entertaining, logical whole. You were the first professional to read the entire draft, and your enthusiasm and encouragement gave me the confidence to get it done.

I have such heartfelt appreciation for my beta readers who gave so generously of their time to help a stranger writing a book. In particular, personal thanks go to:

- Coralie Murray, your thoughtful, enthusiastic conversations brought renewed focus to my message and made the editing process so much more enjoyable.
- Janice Fink, your sentence-level notations and no-nonsense reader approach sharpened my final draft; your input simply made this book better.
- Rusty Williams, amid relaunching your own book, you found time to read mine. Your deep experience in the working and service dog world gave me invaluable insight.
- Petra Brodnik, my Instagram friend, and fellow Havi fan, you and Aiya were the first beta readers to finish, and you did so while on your vacation! Thank you for your unwavering support amid your own very busy life.
- My colleague and friend, Frank Lenox. I know you're not getting a small dog anytime soon, but your scientific knowledge and probing questions helped me tighten up arguments, and your encouragement meant a great deal.

Sincere and humble gratitude goes to the top-level scientists who corresponded with me regarding their work. Your generosity with your time and research made *Downward Sizing Dog* a book that will deepen readers' understanding.

Special appreciation to the small dog owners whose stories helped my research come alive. Your experiences shared here show others what small dog life is about.

And it should go without saying, but it won't: nearly all credit belongs to the dogs who've graced my life. You taught me how to be.

ABOUT KAREN LENA IZZO

Karen Izzo has been owned by fourteen dogs in her short sixty-one years here, and that's not including several black labs, more golden retrievers, and a gorgeous standard poodle who made their (figurative) marks on her life.

Karen's few dogless years were those she spent in school, earning her B.S. in English from Central Michigan University and her law degree from the University of San Francisco. Her early legal career focused on crafting snarky legal briefs and translating complex tax law into understandable English. She eventually discovered the only thing she loved about law was teaching clients to understand the incomprehensible (generation-skipping tax, we're looking at you), so she put away her legal briefs and turned to real education.

For the past twenty-plus years, Karen's taught English in a suburban high school in Rhode Island, where she's also served for over a decade as managing Department Chair. She's been honored with national recognition as a Distinguished Educator through the Coca-Cola Scholars Award. She's been named one of Deb Ruggiero's Amazing Women and received Youth Pride's Angel Award for her writing and work on behalf of LGBTQI youth.

When Karen's not teaching, she's writing, and when she's not writing, she's outdoors in, on, or near the ocean. And almost always, she's accompanied by Phoebe, Scout, and her spouse of thirty-six years. Karen and Ron have three grown children—one in LA, one in Orange County, and one she's hoping will stay nearby on the East Coast.

Downward Sizing Dog: The Companion E-Book

If you're interested in more information and can't wait for the next book, *The Small Dog Rules*, you'll want to download our *Downward Sizing Dog Book Companion*. It's where we're compiling extra information to extend your enjoyment and provide some support if you're looking for or loving a small dog.

This bonus e-book format will enable us to update information easily and allow us to include important resources and lots of photos (wait until you see Phoebe and Scout as puppies…).

The e-book is free for readers of *Downward Sizing Dog: A Reformed Big Dog Snob Defends the Small Dog Life*, and you can access it at:
https://smalldogrules.com/smalldogreads

You can also reach Karen Lena Izzo at:
kizzo@smalldogrules.com
or
kizzo@downwardsizingdog.com

Endnotes

A note about Endnotes. So, my students would be shocked… SHOCKED…to see that I've not followed lock-step with a particular style manual for my Endnotes. But I have a good reason. Or two. First, I hate rules. Okay, that's not the reason (though it's true). In actuality, most of these notes are from Creative Commons License Open Source journals. This relatively new movement allowing free flow of scientific information is truly game-changing for advancing knowledge, and I thought it important to honor that by following the writer's instructions for citing their work.

The other big reason for not stressing over rigid adherence to citation styles is that including a lengthy URL in the print version is just silly. *Downward Sizing Dog* isn't a peer-reviewed paper. Nor do I think most readers are going to snuggle up with my citations and start pouring over dense articles. To me, it is enough for this type of book that a reader so inclined be able to find the original source without a lot of fuss. The following notes allow that, with a little help from Google. In addition, both my website and the e-book will contain hyper-links to the articles available online.

All Creative Commons Licenses are available at https://creativecommons.org/licenses

ENDNOTES

Chapter Two: Run with the Big Dogs, or Stay on the Porch

[1] American Kennel Club. "Most Popular Breeds." Accessed August 12, 2022.

[2] "2 to 20 Years: Boys Stature for Age." National Center for Health Statistics & National Center for Chronic Disease Prevention and Health Promotion. May 30, 2000.

[3] Southwick Associates. "Hunting in America: An Economic Force." PDF. 2018.

[4] M. Cobb *et al*. "The advent of canine performance science: offering a sustainable future for working dogs." Behav. Processes. 2015 Jan; 110:96-104.

[5] Sheila Goffe. "Terrorism Experts, Dog Breeders Gather to Reverse Shortage of Domestic Bomb Sniffing Dogs." *The Hill*. August 21, 2022.

[6] Yolanda F. Bradshaw. "The Impact of Breed Identification, Potential Adopter Perceptions and Demographics, and Dog Behavior on Shelter Dog Adoptability." Master's Thesis, The Ohio State University. 2021.

[7] Bradshaw. "The Impact of Breed Identification…".

[8] Packaged Facts Pet Owners Survey 2015 as reported in "Ownership of Small Dogs on the Rise." Petfoodindustry.com. January 31, 2015.

[9] Roberto A. Ferdman. "Tiny Dogs are Taking Over this Country." *The Washington Post*. Feb. 26, 2015.

[10] AKC Staff. "Top Ten Dog Breeds of the 1930s." *American Kennel Club*. February 13, 2015.

[11] AKC Staff. "Most Popular Dog Breeds – Full Ranking List." *American Kennel Club*. Mar. 28, 2019.

[12] Ferdman. "Tiny Dogs...".

[13] Megan S. Broadway. "Are Large Dogs Smarter Than Small Dogs? Investigating Within Species Differences in Large and Small Dogs: Spatial Memory." Master's Thesis, University of Southern Mississippi. 2015.

[14] Ghirlanda S, Acerbi A, Herzog H, Serpell JA (2013) "Fashion vs. Function in Cultural Evolution: The Case of Dog Breed Popularity." PLoS ONE 8(9): e74770. Creative Commons Attribution License.

Chapter Three: Small Dogs Are Dogs

[15] Anders Bergström, D.W.G. Stanton, U.H. Taron, *et al*. "Grey Wolf Genomic History Reveals A Dual Ancestry Of Dogs." *Nature* 607, 313–320 (2022). Creative Commons Attribution 4.0 International (CC BY 4.0).

Endnotes

[16] Linda Y. Rutledge and Paul J. Wilson. "Considering Pleistocene North American Wolves and Coyotes in the Eastern Canis Origin Story." *Ecology and Evolution.* Jun 5, 2021. Creative Commons Attribution 4.0 International (CC BY4.0).

[17] Freedman AH, Gronau I, Schweizer RM, Ortega-Del Vecchyo D, Han E, Silva PM, *et al.* 2014. "Genome Sequencing Highlights the Dynamic Early History of Dogs." *PLoS Genet* 10(1): e1004016. Creative Commons CC0 Public Domain Dedication.

[18] Gray, M.M., Sutter, N.B., Ostrander, E.A. *et al.* "The *IGF1* Small dog Haplotype is Derived from Middle Eastern Grey Wolves." *BMC Biol* 8, 16. 2010. Creative Commons Attribution 2.0 International (CC BY 2.0).

[19] Klütsch, C.F., de Caprona, M.D.C. "The IGF1 Small Dog Haplotype is Derived from Middle Eastern Grey Wolves: a Closer Look at Statistics, Sampling, and the Alleged Middle Eastern Origin of Small Dogs." *BMC Biol* 8, 119 (2010). Creative Commons Attribution 2.0 International (CC BY 2.0).

[20] Elaine A Ostrander, et al. "Dog10k: An International Sequencing Effort To Advance Studies Of Canine Domestication, Phenotypes And Health." *National Science Review*, Volume 6, Issue 4, July 2019, Pages 810–824.. Creative Commons Attribution License (CC BY).

[21] J. Plassais, J. Kim, B.W. Davis, *et al.* "Whole genome sequencing of canids reveals genomic regions under selection and variants influencing morphology." *Nat Commun* 10, 1489 (2019). Creative Commons Attribution License (CC BY 4.0).

[22] "Why Small Dogs are Small." UNews Archive. University of Utah. March 25, 2011.

[23] "Why Small Dogs…".

[24] A.R. Boyko, P. Quignon, L. Li, J.J. Schoenebeck, J.D. Degenhardt, K.E. Lohmueller, et al. (2010). "A Simple Genetic Architecture Underlies Morphological Variation in Dogs." *PLoS Biol* 8(8): e1000451. Creative Commons CC0 Public Domain Dedication.

[25] Alex Fox. "Mutation That Gave Us Tiny Dogs Found in Ancient Wolves." *Smithsonian Magazine*. January 27, 2022.

[26] "Tiny Chihuahua Saves Life Of 1-Year-Old Colo. Boy Attacked By Rattlesnake." *The Seattle Times*. July 22, 2007.

[27] "Heroic 10-Pound Dog Saves Little Girl From Venomous Rattlesnake In Texas." *The Huffington Post*. August 14, 2013. (The snake struck Psycho in the eye, but thankfully the pup received anti-venom treatment in time.)

[28] Ravindra Seshu. "Pet Dog Dies Saving Owner From Snake Bite." *Deccan Chronicle*. April 13, 2020.

[29] "Dog Suffers Bite While Protecting Owner From Rattlesnake." *Inquirer.net*. August 6, 2020.

[30] Lynda Baquero. "Dog Saves Connecticut Family from House Fire by Alerting Owners to Danger." NBC New York. September 29, 2020.

[31] "Heroic Dog Saves Bradenton Family From House Fire." WFLA News Channel 8. September 17, 2019.

[32] "Hero Dog Saves Family from Fire." CBS News. October 5, 2016.

[33] Terry Camp. "Tuscola County family credits their dog's strange behavior with alerting them to house fire." ABC 12 News. March 22, 2022.

[34] Paul Cox. "Smoky The Terrier: A Tiny War Hero Immortalized." Honey, Stop the Car. NPR. August 3, 2011.

[35] William A. Wynne. *Yorkie Doodle Dandy: A Memoir, 7th Edition*. Mansfield, OH. Wynnesome Press, Ltd., 1996.

[36] Mindy Norton. "Rags—World War I Hero Dog." Alabama Public Radio. May 29, 2021. Rags' story is recounted in Grant Hayter-Menzies' book *From Stray Dog to WWI Hero*. Potomac Books, 2015.

[37] A. Ortal, A. Rodríguez, M.P. Solis-Hernández, *et al.* "Proof of Concept for the Use of Trained Sniffer Dogs to Detect Osteosarcoma." *Sci Rep* 12, 6911. 2022. Creative Commons Attribution License (CC BY 4.0).

[38] Pearl Tesler. "Dogs Smell Time." Spectrum Blog. Exploratorium.edu. June 19, 2014.

[39] Experimental Biology. "Study Shows Dogs Can Accurately Sniff Out Cancer in Blood: Canine Cancer Detection Could Lead to New Noninvasive, Inexpensive Ways to Detect Cancer." *ScienceDaily*. 8 April 2019.

[40] Heather Spaulding. "Smelling A Cure for Parkinson's." *The Island's Sounder*. November 23, 2019.

[41] Lourdes Perez. "FIU is Training Dogs to Detect COVID-19." Campus and Community. Florida International University. January 22, 2021

[42] "You Did What With Your Weiner." https://youdidwhatwithyourweiner.com/

Chapter Four: Small Dogs Are Smart

[43] Joshua J. Mark. "Dogs In The Ancient World." *World History Encyclopedia*. January 14, 2019.

Endnotes

[44] Nuno Henrique Franco. "Animal Experiments in Biomedical Research: A Historical Perspective." *Animals* vol. 3,1 23873. 19 Mar. 2013. Creative Commons Attribution License (CC BY 4.0).

[45] Franco. "Animal Experiments...".

[46] Roberta Kalechofsky. "Dedicated to Descartes' Niece: The Women's Movement In the Nineteenth Century and Anti-Vivisection." *Between The Species. San Francisco Bay Institute Quarterly Journal.* Spring 1992.

[47] Kalechofsky. "Dedicated to Descartes' Niece...".

[48] "Animals." *Project Gutenberg's Voltaire's Philosophical Dictionary.* E-book Release Date: 2006. Original text 1764.

[49] George J. Romanes. *Animal Intelligence.* New York. D. Appleton and Company. 1884.

[50] Edward L. Thorndike. *Animal Intelligence: An Experimental Study of the Associative Process in Animals.* New York. The MacMillon Company. 1898. Digitized by Internet Archive.

[51] "Fellows." The Royal Society of Canada. Accessed October 25, 2022.

[52] Selby Frame. "Adventures in Dog Research with Stanley Coren." Members. American Psychological Association. April 25, 2017. Accessed October 25, 2022. Fair Use Exception per APA Guidelines.

[53] Frame. "Adventures in Dog Research with Stanley Coren."

[54] The Horowitz Dog Cognition Lab at Barnard College, Yale University.

[55] Canine Cognition Center at Yale.

[56] The Canine Science Collaboratory (Wynne).

[57] Duke Canine Cognition Center.

[58] Canine Cognition and Human Interaction Lab. University of Nebraska-Lincoln.

[59] Canine Olfaction Lab. Texas Tech University.

[60] Arizona Canine Cognition Center. The University of Arizona

[61] Clever Canine Lab. University of Auckland.

[62] MacLean EL, Fine A, Herzog H, Strauss E and Cobb ML (2021) "The New Era of Canine Science: Reshaping Our Relationships With Dogs." *Front. Vet. Sci.* 8:675782. Creative Commons Attribution License (CC BY).

[63] Megan Broadway. "Are Large Dogs Smarter Than Small Dogs? Investigating Within Species Differences in Large and Small Dogs: Spatial Memory." Master's Thesis. The University of Southern Mississippi. (2015).

[64] D.J. Horschler, B. Hare, J. Call, et al. "Absolute Brain Size Predicts Dog Breed Differences In Executive Function." *Anim Cogn* 22, 187–198 (2019).

[65] M. Foraita, T. Howell, & P. Bennett, "Development of the Dog Executive Function Scale (DEFS) for Adult Dogs." *Anim Cogn* (2022). Creative Commons Attribution License (CC BY 4.0).

[66] Megan S. Broadway et al. "Does Size Really Matter? Investigating Cognitive Differences in Spatial Memory Ability Based on Size in Domestic Dogs." ScienceDirect. *Behavioural Processes*. Volume 138. 2017. Pages 7–14.

[67] Erin E. Hecht et al. "Significant Neuroanatomical Variation Among Domestic Dog Breeds." *Journal of Neuroscience*. September 25, 2019. Creative Commons Attribution License (CC BY 4.0).

[68] Mark Rishniw and Curtis W. Dewey. "Little Brainiacs and Big Dummies: Are We Selecting for Stupid, Stout, or Small Dogs?" *Open Vet Journal*. 2021 Jan-Mar; 11(1): 107–111. Published online Feb 6, 2021. Open Access under African Journals Online.

[69] Rishniw and Dewey. "Little Brainiacs and Big Dummies."

[70] Horschler et al. "Absolute Brain Size Predicts…".

[71] Parson Russell Terrier Association of America.

[72] Dr. Stanley Coren. "Can A Dog's Size Predict Its Intelligence?" *Canine Corner*. *Psychology Today*. August 31, 2016.

Chapter Five: Small Dogs Are Powerful Health Boosters

[73] Maria Petersson, Kerstin Uvnäs-Moberg, Anne Nilsson, Lise-Lotte Gustafson, Eva Hydbring-Sandberg, and Linda Handlin. "Oxytocin and Cortisol Levels in Dog Owners and Their Dogs Are Associated with Behavioral Patterns: An Exploratory Study". *Frontiers in Psychology*. Published online 2017 Oct 13. Open Access Under Creative Commons License (CC BY).

[74] Petersonn et al. "Oxytocin and Cortisol Levels…".

[75] Wirobski, G., Range, F., Schaebs, F.S. et al. "Life Experience Rather Than Domestication Accounts for Dogs' Increased Oxytocin Release During Social Contact with Humans." *Sci Rep* 11, 14423 (2021). Open Access Under Creative Commons License (CC BY 4.0).

[76] Gee NR, Rodriguez KE, Fine AH and Trammell JP (2021). "Dogs Supporting Human Health and Well-Being: A Biopsychosocial Approach." *Front. Vet. Sci.* 8:630465.

[77] Ann Robinson. "Dogs Have A Magic Effect: How Pets Can Improve Our Mental Health". *The Guardian*. March 17, 2020. Accessed October 27, 2022.

Endnotes

[78]John Polheber, et al. "The Presence of a Dog Attenuates Cortisol and Heart Rate in the Trier Social Stress Test Compared to Human Friends." *J Behav Med.* 2014 Oct; 37(5):860-7.

[79] Machová, K.; Procházková, R.; Vadroňová, M.; Součková, M.; Prouzová, E. "Effect of Dog Presence on Stress Levels in Students under Psychological Strain: A Pilot Study." *Int. J. Environ. Res. Public Health* 2020, *17*, 2286. Creative Commons Attribution License (CC BY).

[80]Christy L. Hoffman, *et al.* An Examination of Adult Women's Sleep Patterns and Sleep Routines In Relation to Pet Ownership and Bed Sharing. Anthrozoos. Volume 31, 2018. Issue 6.

[81] Francesca Moretti *et al.* "Pet Therapy In Elderly Patients With Mental Illness." *Psychogeriatrics*, June 2011. Free Access Journal.

[82] Vegue Parra, Eva, Jose Manuel Hernández Garre, and Paloma Echevarría Pérez. 2021. "Benefits of Dog-Assisted Therapy in Patients with Dementia Residing in Aged Care Centers in Spain." *International Journal of Environmental Research and Public Health* 18, no. 4: 1471. Creative Commons Attribution License (CC BY).

[83] Rodrigo-Claverol, Maylos, Belén Malla-Clua, Carme Marquilles-Bonet, Joaquim Sol, Júlia Jové-Naval, Meritxell Sole-Pujol, and Marta Ortega-Bravo. 2020. "Animal-Assisted Therapy Improves Communication and Mobility among Institutionalized People with Cognitive Impairment." *International Journal of Environmental Research and Public Health* 17, no. 16: 5899. Creative Commons Attribution License (CC BY).

[84] Quintavalla, F.; Cao, S.; Spinelli, D.; Caffarra, P.; Rossi, F.M.; Basini, G.; Sabbioni, A. "Effects of Dog-Assisted Therapies on Cognitive Mnemonic Capabilities in People Affected by Alzheimer's Disease." *Animals* 2021, *11*, 1366. Creative Commons Attribution License (CC BY).

[85] Cotoc, C.; Notaro, S. "Race, Zoonoses and Animal Assisted Interventions in Pediatric Cancer." *Int. J. Environ. Res. Public Health* 2022, *19*, 7772. Creative Commons Attribution License (CC BY).

[86] "Pet Therapy." Penn State Cancer Institute. Accessed October 28, 2022.

[87] "Caring Canines." Memorial Sloan-Kettering Cancer Center. Accessed October 28, 2022.

[88] "Caring Canines." Mayo Clinic. Accessed October 28, 2022.

[89] American Veterinary Medical Association. *Animal Assisted Interventions: Guidelines.* Accessed October 28, 2022.

[90] Meints K, Brelsford VL, Dimolareva M, Maréchal L, Pennington K, Rowan E, et al. (2022) "Can Dogs Reduce Stress Levels In School Children? Effects Of Dog-Assisted Interventions On Salivary Cortisol In Children With And

Without Special Educational Needs Using Randomized Controlled Trials." PLoS ONE 17(6): e0269333. Creative Commons Attribution License (CC BY).

[91] "Service Animal Misconceptions." *ADA National*. https://adata.org. Accessed October 27, 2022.

[92] "Post-Traumatic Stress Disorder (PTSD)." Patient Care and Health Information. Mayo Clinic. Accessed October 28, 2022.

[93] Lucy Laing. "It's A Dog's Life! Stressed High Level Executives Are Sent To A Pooch Clinic Where Psychologists Discover The Havanese Breed Is The Best For Reducing Anxiety." *The Daily Mail*. October 20, 2017.

[94] Acquadro Maran, D.; Capitanelli, I.; Cortese, C.G.; Ilesanmi, O.S.; Gianino, M.M.; Chirico, F. "Animal-Assisted Intervention and Health Care Workers' Psychological Health: A Systematic Review of the Literature." *Animals* 2022, *12*, 383. Open Access through MDPI.

[95] "Meet Some of the 8,000 Dogs That 'Work' at Amazon." AboutAmazon.com. February 2, 2022.

[96] "Meet Some of the Dogs…."

[97] Sophie Susannah Hall and Daniel Simon Mills. "Taking Dogs Into the Office: A Novel Strategy for Promoting Work Engagement, Commitment and Quality of Life." *Front Vet Sci*. 2019; 6: 138. Published online 2019 May 7. Creative Commons Attribution License (CC BY).

[98] Elisa Wagner and Miguel Cunha. "Dogs at the Workplace: A Multiple Case Study." *Animals*. January 5, 2021. Creative Commons Attribution License (CC BY).

[99] Molly McHugh-Johnson. "Working from Home is Ruff. Dooglers Make It Better." *The Keyword. Inside Google*. October 16, 2020.

[100] Jordan Smith. "Why You Should Be Pro Puppy In the Workplace." *Inc*.com. July 23, 2014.

Chapter Six: Small Dogs Are Easier on Our Health, Our Homes, and Our Planet

[101] "Allergens: Animals." *Health Encyclopedia*. University of Rochester Medical Center. Accessed October 28, 2022.

[102] "Allergic to Your Dog? Easy Tips to Prevent and Control Your Allergy." Veterinary Public Health Program. The Ohio State University College of Veterinary Medicine. Accessed October 28, 2022.

[103] Customs Duties. US Code 19 (2011) § 1308 - Prohibition on importation of dog and cat fur products.

[104] Tom Usher. "Your Dog Is Ruining The Environment." *Vice.com*. May 2018.

Endnotes

[105] Gregory S. Okin. "Environmental Impacts of Food Consumption of Dogs and Cats." PLOS One. August, 2017. Creative Commons Attribution License.

[106] Ri'An Jackson. "Pooch Poop Poses Peril." *Spartan News Room*. Michigan State University School of Journalism. March 13, 2020.

[107] "Pet Waste." Department of Health. State of Rhode Island. Accessed October 28, 2022.

[108] Jocelyn Timperley. "Should We Give Up Flying for the Sake of the Environment?" *Smart Guide to Climate Change*. BBC. February 18, 2020.

[109] "Combat Climate Change." Sustainable Travel International. 2020. Accessed October 28, 2022.

[110] "Air Travel Consumer Report." Office of Aviation Enforcement and Proceedings, Aviation Consumer Protection Division, United Stated Department of Transportation. February 2019.

Chapter Eight: Small Dogs Are Less Expense (Unless You Spoil Them, but That's on You)

[111] "How Much Does It Cost To Have a Dog?" ASPCA Pet Health Insurance. Accessed October 28, 2022.

[112] Amy Danise. "Most Expensive Dog Breeds for Insurance." *Forbes Advisor*. Updated October 1, 2022.

[113] "Pet Industry Market Size, Trends, and Ownership Statistics." American Pet Products Association. Accessed October 28, 2022.

[114] "Pet Food Market Size Share…Regional Forecast 2022–2029." Animal Nutrition/Pet Food Market. *Fortune Business Insights*. Accessed October 28, 2022.

[115] "Pet Toys Market Size Share…Regional Forecast 2022–2029." Consumer Goods/Pet Toys Market. *Fortune Business Insights*. Accessed October 28, 2022.

[116] Jamie Farkas. "The Top 30 Most Expensive Dog Breeds." *Save Money At Home*. Go Banking Rate. September 6, 2022.

[117] "Pet Industry Market Size, Trends, and Ownership Statistics."

[118] AVMA Veterinary Economics Division. "Dog Ownership and Veterinary Visits by Income Bracket." *JAVMA News*. January 15, 2020.

Chapter Nine: Small Dogs Live Longer

[119] Urfer, S.R., Kaeberlein, M., Promislow, D.E.L. *et al.* "Lifespan of Companion Dogs Seen in Three Independent Primary Care Veterinary Clinics in the United States." *Canine Genet Epidemiol* 7, 7 (2020). Creative Commons International License (CC BY 4.0).

[120] Paul S. Brookes and Ana Gabriela Jimenez. "Metabolomics of Aging in Primary Fibroblasts from Small and Large Breed Dogs." *GeroScience* 43, June 16, 2021. Creative Commons International (CC BY 4.0).

[121] "What is One Health?" *One Health Commission*. Accessed October 29, 2022.

Chapter Ten: The Small Dog Life Fits Every Life Stage

[122] Lenkei, R., Faragó, T., Zsilák, B. *et al.* "Dogs (*Canis Familiaris*) Recognize Their Own Body As A Physical Obstacle." *Sci Rep* 11, 2761 (2021). Creative Commons License (CC BY 4.0).

[123] Pet Population and Ownership Trends in the United States, 5th Ed. *YouTube*. https://youtu.be/7vWsNqTO_Vk Accessed October 28, 2022.

The Case for Small Dogs and Kids

[124] Meints K, Brelsford VL, Dimolareva M, Maréchal L, Pennington K, Rowan E, et al. (2022) "Can Dogs Reduce Stress Levels in School Children? Effects of Dog-assisted Interventions on Salivary Cortisol in Children with and without Special Educational Needs Using Randomized Controlled Trials." PLoS ONE 17(6): e0269333. Creative Commons Attribution License (CC BY 4.0).

[125] Jana Meixner and Kurt Kotrschal (2022) "Animal-Assisted Interventions With Dogs in Special Education—A Systematic Review." *Front. Psychol.* 13:876290. Creative Commons Attribution License. (CC BY 4.0).

[126] Jill Steel *et al.* "Reading to Dogs in School: An Exploratory Study of Teacher Perspectives." EDUCATIONAL RESEARCH 2021, VOL. 63, NO. 3, 279-301 Open Access through Routledge Taylor & Francis Group.

[127] Antonio Benítez-Burraco, Daniela Pörtl, and Christoph Jung. "Did Dog Domestication Contribute to Language Evolution?" *Front Psychol*. 2021; 12: 695116. Creative Commons Attribution License (CC BY 4.0).

[128] Purewal, R.; Christley, R.; Kordas, K.; Joinson, C.; Meints, K.; Gee, N.; Westgarth, C. "Companion Animals and Child/Adolescent Development: A Systematic Review of the Evidence." *Int. J. Environ. Res. Public Health* 2017, *14*, 234. Creative Commons Attribution License.

[129] Gretchen Carlisle. "The Social Skills and Attachment to Dogs of Children with Autism Spectrum Disorder." *J Autism Dev Disord* 45, 1137–1145 (2015).

[130] Dollion N, Grandgeorge M, Saint-Amour D, Hosein Poitras Loewen A, François N, Fontaine NMG, Champagne N and Plusquellec P (2022) "Emotion Facial Processing in Children With Autism Spectrum Disorder: A Pilot Study of the Impact of Service Dogs." Front. Psychol. 13:869452. Creative Commons Attribution License (CC BY 4.0).

[131] Ferran Marsa-Sambola *et al.* "Quality of Life and Adolescents' Communication with their Significant Others (mother, father, and best friend):

Endnotes

The Mediating Effect of Attachment to Pets." *Attachment & Human Development*. February 20, 2017. 19:3, 278-297.

[132] Minatoya, M.; Araki, A.; Miyashita, C.; Itoh, S.; Kobayashi, S.; Yamazaki, K.; Ait Bamai, Y.; Saijyo, Y.; Ito, Y.; Kishi, R. "Cat and Dog Ownership in Early Life and Infant Development: A Prospective Birth Cohort Study of Japan Environment and Children's Study." *Int J Environ Res Public Health*. 2020 Jan; 17(1): 205. Published online 2019 Dec 27. Creative Commons Attribution License (CC BY 4.0).

[133] Carri Westgarth et al. "Dog Owners Are More Likely To Meet Physical Activity Guidelines Than People Without a Dog: An Investigation Of The Association Between Dog Ownership And Physical Activity Levels In a UK Community." Sci Rep. 2019 Apr 18;9(1):5704. Creative Commons Attribution License (CC BY 4.0).

[134] Hielscher-Zdzieblik, B.; Froboese, I.; Serpell, J.; Gansloßer, U. "Impact of Dog's Age and Breed on Dog Owner's Physical Activity: A German Longitudinal Study." *Animals* 2022, *12*, 1314. Creative Commons Attributions License (CC BY 4.0.)

[135] "The Role of Breed in Dog Bite Risk and Prevention." American Veterinary Medicine Association. April 12, 2012. PDF Accessed October 29, 2022.

[136] "The Role of Breed…".

[137] Tony Reynolds. "New Study Identifies Most Damaging Dog Bites By Breed." Newstat. American Animal Hospital Association Publication. June 6, 2019.

[138] Lisa Maria Glenk and Sandra Foltin. "Therapy Dog Welfare Revisited: A Review of the Literature." *Vet Sci*. 2021 Oct; 8(10): 226. Open Access under MDPI policy.

[139] Sophie S. Hall, Beverley J. Brown, and Daniel S. Mills. "Developing and Assessing the Validity of a Scale to Assess Pet Dog Quality of Life: Lincoln P-QoL." *Front Vet Sci*. 2019; 6: 326. Published online 2019 Sep 26. Creative Commons Attribution License (CC BY 4.0).

The Case for Small Dogs and Millennials

[140] Steve Dale. "Millennials Lead The Way In Pet Care." Fear Free Pets Online. Accessed October 29, 2022.

[141] "Seventy-Six Percent of Millennials are Pet Parents. Here's What They've Been Buying." YPulse. August 24, 2020.

[142] "Seventy-Six Percent…".

[143] Olivia Ledbetter. "Pets vs. Parenthood: Why Millennials Are Owning Pets Instead of Having Kids." *Millennial Marketing*. Accessed October 29, 2022.

The Case for Small Dogs and Boomers

[144] "Report: Pet Marketers Might Want To Focus More On Baby Boomers." *Pet Product News*. December 9, 2019.

[145] Jared Ortaliza, Krutica Amin, and Cynthia Cox. "COVID-19 Leading Cause of Death Ranking." March 24, 2022.

[146] Christina Cheakalos. "10 Reasons to Get a Dog When You're Over 50." AARP. Visited May 28, 2022.

[147] Surma, S., Oparil, S. & Narkiewicz, K. "Pet Ownership and the Risk of Arterial Hypertension and Cardiovascular Disease." *Curr Hypertens Rep* (2022). Creative Commons International License (CC BY 4.0).

[148] Caroline K. Kramer, Sadia Mehmood, and Renée S. Suen. "Dog Ownership and Survival: A Systematic Review and Meta-Analysis." *Circulation: Cardiovascular Quality and Outcomes*. Oct 8, 2019. Free Access article.

[149] Mwenya Mubanga, Liisa Byberg, Agneta Egenvall, Erik Ingelsson, Tove Fall. "Dog Ownership and Survival After a Major Cardiovascular Event." *Circulation: Cardiovascular Quality and Outcomes*. October 8, 2019. Free access article.

[150] Jacob Ausubel. "Older People Are More Likely To Live Alone In The US Than Elsewhere In The World." March 10, 2020. Pew Research Center.

[151] "Baby Boomer Report." US News Market Insights. 2015. Accessed October 29, 2022.

[152] "Boomers Have Big Travel Plans in 2020." *Groups Today*. February 6, 2020.

[153] Vicky Levy. "Despite COVID-19 Concerns, Many Boomers Plan to Travel in 2021." *Life and Leisure*. AARP. April 2021.

The Case for Small Dogs and Pet Parents in Their Seventies and Beyond

[154] R. Scheibeck *et al.* "Elderly People In Many Respects Benefit from Interactions with Dogs." *Eur J Med Res*. December 2, 2011.

[155] Nataša Obradović *et al.* "Understanding the Benefits, Challenges, and the Role of Pet Ownership in the Daily Lives of Community-Dwelling Older Adults: A Case Study." *Animals*. September 7, 2021. Creative Commons Attribution License (CC BY 4.0).

[156] Taniguchi Y, Seino S, Headey B, Hata T, Ikeuchi T, Abe T, et al. "Evidence That Dog Ownership Protects Against the Onset of Disability in an Older Community-Dwelling Japanese Population." PLOS ONE. February 23, 2022. Creative Commons Attribution License (CC BY 4.0).

[157] Westgarth C, Christley RM, Jewell C, German AJ, Boddy LM, Christian HE. "Dog Owners Are More Likely To Meet Physical Activity Guidelines Than People Without a Dog: An Investigation Of The Association Between Dog

Endnotes

Ownership And Physical Activity Levels In a UK Community." *Scientific Reports*. April 18, 2019. Creative Commons International License (CC BY 4.0).

[158] Clement Meier and Jurgen Maurer. "Buddy or burden? Patterns, perceptions, and experiences of pet ownership among older adults in Switzerland." *European Journal of Ageing*, 1-12 – April. Creative Commons International License (CC BY 4.0).

[159] David G. Pearson and Tony Craig. "The Great Outdoors? Exploring the Mental Health Benefits of Natural Environments." Frontiers in Psychology. October 21, 2014. Creative Commons Attribution License (CC BY 4.0).

[160] Si-Jie Li, Yu-Feng Luo, Zi-Chuan Liu, Lei Xiong, Bo-Wei Zhu "Exploring Strategies for Improving Green Open Spaces in Old Downtown Residential Communities from the Perspective of Public Health to Enhance the Health and Well-Being of the Aged." *Journal of Healthcare Engineering*, vol. 2021, Article ID 5547749, 2021. Creative Commons Attribution License.

[161] Wood L, Martin K, Christian H, Nathan A, Lauritsen C, Houghton S, *et al*. "The Pet Factor—Companion Animals as a Conduit for Getting to Know People, Friendship Formation and Social Support." PloS ONE 10(4): e0122085. April 29, 2015. Creative Commons Attribution License.

[162] Katie Potter, Hachem Saddiki, and Laura B. Balzer. "Dog Walking Mediates The Relationship Between Dog Ownership And Neighborhood Social Interaction." *Innovations in Aging*. November 8, 2019. Creative Commons CC BY License.

[163] Marcello Siniscalchi *et al*. "Lateralized Behavior And Cardiac Activity Of Dogs In Response To Human Emotional Vocalizations." *Sci Rep* 8, 77 (2018). Creative Commons Attribution License (CC BY 4.0).

[164] Zafra-Tanaka, J.H., Pacheco-Barrios, K., Tellez, W.A. *et al*. "Effects of dog-assisted therapy in adults with dementia: a systematic review and meta-analysis." *BMC Psychiatry* 19, 41 (2019). Creative Commons Attribution International License (CC BY 4.0).

[165] Merritt Whitley. "Furry Friends Welcome: A Guide To Pet-Friendly Senior Living." April 13, 2021. *Caregiver Resources*. A Place for Mom.

[166] Billy Francis. "Enjoy Your Golden Years With Your Goldie at These Assisted Living Facilities." Pet Travel Blog. BringFido.com. February 12, 2020.

[167] Mark dos Anjos. "The Five Best Dog Breeds for Senior Citizens." Pethelpful.com. April 21, 2022.

[168] J. A. Stevens. "Nonfatal Fall-Related Injuries Associated with Dogs and Cats." Morbidity and Mortality Weekly Report. Centers for Disease Control and Prevention. March 27, 2009.

[169] "Non-fatal Fall-Related Injuries…".

[170] "Nonfatal Fall-Related Injuries...".

At Every Age: Emergency and Estate Planning for Small Dogs

[171] James T. Brennan, *Bequests for the Care of Specific Animals*, 6 Duq. L. Rev. 15 (1967). Available at: https://dsc.duq.edu/dlr/vol6/iss1/3

[172] Gerry W. Beyer. "Pet Animals: What Happens When Their Humans Die?" 40 SANCLR 617 (2000). Reprinted online with permission at https://www.animallaw.info/article/wills-trusts-pet-animals-what-happens-when-their-humans-die

[173] Map of States with Companion Animal (Pet) Trust Laws. Michigan State University Animal Legal and Historical Center. https://www.animallaw.info/content/map-states-companion-animal-pet-trust-laws. Accessed October 29, 2022.

[174] Map of States with Companion Animal (Pet) Trust Laws.

[175] "Providing For Your Pets In the Event of Your Death Or Hospitalization." Committee on Animal Law. New York City Bar Association. May 2016. Accessed October 29, 2022.

[176] "Providing For Your Pets...".

Chapter Eleven: But Wait. What About Small Dog Syndrome?

[177] Ken Layne. "Eradicate Small Dogs Now." *Gawker*. August 12, 2013.

[178] Anderson, K.H.; Yao, Y.; Perry, P.J.; Albright, J.D.; Houpt, K.A. 2022. "Case Distribution, Sources, and Breeds of Dogs Presenting to a Veterinary Behavior Clinic in the United States from 1997 to 2017" *Animals* 12, no. 5: 576. Creative Commons Attribution License.

[179] Mikkola, S., Salonen, M., Puurunen, J. *et al.* "Aggressive Behavior is Affected by Demographic, Environmental and Behavioural Factors in Purebred Dogs." *Nature*. May 3, 2021. Creative Commons Attribution License (CC BY 4.0).

[180] Borgi M, Cogliati-Dezza I, Brelsford V, Meints K and Cirulli F. (2014) "Baby Schema In Human And Animal Faces Induces Cuteness Perception And Gaze Allocation In Children." Front. Psychol. 5:411. Creative Commons Attribution License (CC-BY).

[181] Prokop P, Zvaríková M, Zvarík M, Pazda A and Fedor P. (2021) "The Effect of Animal Bipedal Posture on Perceived Cuteness, Fear, and Willingness to Protect Them." Front. Ecol. Evol. 9:681241. Creative Commons Attribution License (CC-BY).

[182] "The Crazy Nastya** Honey Badger." *YouTube*. Accessed May 30, 2022. https://youtu.be/4r7wHMg5Yjg

Endnotes

[183] McGreevy PD, Georgevsky D, Carrasco J, Valenzuela M, Duffy DL, *et al.* (2013) "Dog Behavior Co-Varies with Height, Bodyweight and Skull Shape." PLoS ONE 8(12): e80529. Open Access Under Creative Commons Attribution License.

[184] Salla Mikkola. "Aggressive Behavior...".

Chapter Twelve: A Brief Bit of Advice About Finding Your Small Dog

[185] Powell, L., Reinhard, C., Satriale, D. *et al.* "Characterizing Unsuccessful Animal Adoptions: Age And Breed Predict The Likelihood Of Return, Reasons For Return And Post-Return Outcomes." *Sci Rep* 11, 8018 (2021). Creative Commons Attribution International License (CC BY 4.0).

[186] Natalia Ribeiro Santos, Alexandra Beck, Cindy Maenhoudt, Charlotte Billy, and Alain Fontbonne. 2021. "Profile of Dogs' Breeders and Their Considerations on Female Reproduction, Maternal Care and the Peripartum Stress—An International Survey." *Animals* 11, no. 8: 2372. Creative Commons Attribution International License (CC BY 4.0).

[187] Shan S, Xu F and Brenig B. (2021) "Genome-Wide Association Studies Reveal Neurological Genes for Dog Herding, Predation, Temperament, and Trainability Traits." *Front. Vet. Sci.* 8:693290. Open Access under Creative Commons CC-BY License. See also Katherine H. Anderson. "Case Distribution…".

[188] Kathleen Morrill *et al.* "Ancestry-Inclusive Dog Genomics Challenges Popular Breed Stereotypes." *Science*. April 29, 2022.

Made in the USA
Las Vegas, NV
30 March 2025